Liaison Functions 2016

Core Programming Area at the 2016 AIChE Annual Meeting

San Francisco, California, USA
13-18 November 2016

ISBN: 978-1-5108-3460-6

Printed from e-media with permission by:

Curran Associates, Inc.
57 Morehouse Lane
Red Hook, NY 12571

proceedings.com

Some format issues inherent in the e-media version may also appear in this print version.

Printed by Curran Associates, Inc. (2017)

AIChE
120 Wall Street, FL 23
New York, NY 10005-4020

Phone: (800) 242-4363
Fax: (203) 775-5177

www.aiche.org

Additional copies of this publication are available from:

Curran Associates, Inc.
57 Morehouse Lane
Red Hook, NY 12571 USA
Phone: 845-758-0400
Fax: 845-758-2633
Email: curran@proceedings.com
Web: www.proceedings.com

TABLE OF CONTENTS

472251 Viability of Low Pressure Ammonia Synthesis Via Reactive Separation

Abstract:

Viability of Low Pressure Ammonia Synthesis via Reactive Separation

Mahdi Malmali, Alon McCormick, and Edward L. CusslerDepartment of Chemical Engineering and Materials Science, University of Minnesota, Minneapolis, MN 55455

Ammonia is a crucial chemical primarily used as an agricultural fertilizer; it is the second most manufactured chemical in the world with more than 160 million tons produced per year[1]. Ammonia is produced via the Haber-Bosch process, where hydrogen (from thermochemical steam-reforming of methane) and nitrogen are combined and react over a promoted iron-based catalyst at temperatures ranging from 380° to 520° C and pressures from 150 to 300 bar[2]. Reaction at such an elevated pressures only achieves partial conversions (typically less than 30%), which necessitates the recycle of unreacted gas using costly phase separation processes. After more than a century of research and optimization, this process is still highly capital- and energy-intensive[3], hence only economically viable in large-scale centralized plants.

We seek to develop technologies for small-scale synthesis of ammonia with zero-carbon footprint[4,5]. This process is powered by electricity produced from stranded intermittent wind or solar energy and can make hydrogen from electrolysis of water and nitrogen from pressure swing adsorption of air. A major drawback for widespread application of such a process is the high cost of the synthesized ammonia. The pressure in this process can be reduced to improve the economics of the process. This is achieved by simultaneous removal of the ammonia from the reaction system and circumventing the reaction equilibrium limitation.

We will initially present the ammonia reaction rate kinetics at low pressures. Then we discuss our findings of the reaction-separation process, which consists of a catalytic reactor and absorber unit; we report the absorption of ammonia on $CaCl_2$, which is utilized to selectively remove the ammonia from the system at high temperatures. Our experimental results demonstrate the high capacity of the absorption material at close-to-reaction temperature, which suppresses the equilibrium and increases production rates in our reaction-separation system; conversions as high as 80% at significantly reduced pressure (< 20 bar) are achieved. The effect of different operating conditions, including reaction temperature, reaction pressure, absorption temperature, gas transport, etc. on the kinetics of ammonia synthesis will be discussed in detail. For proof-of-concept, we compare the results obtained from high pressure reaction and low pressure reaction-separation process.

(1) http://minerals.usgs.gov/minerals/pubs/commodity/nitrogen/mcs-2011-nitro.pdf. (2) *Catalytic Ammonia Synthesis: Fundamentals and Practice*; Springer Science & Business Media, 2013. (3) Vojvodic, A.; Medford, A. J.; Studt, F.; Abild-Pedersen, F.; Khan, T. S.; Bligaard, T.; Nørskov, J. K. Exploring the Limits: A Low-Pressure, Low-Temperature Haberâ€"Bosch Process. *Chem. Phys. Lett.* **2014**, *598*, 108.

Â (4) Reese, M.; Marquart, C.; Malmali, M.; Wagner, K.; Buchanan, E.; McCormick, A.; Cussler, E. L. Performance of a Small-Scale Haber Process. *Ind. Eng. Chem. Res.* **2016**, *55*(13), 3742.

Â (5) Himstedt, H. H.; Huberty, M. S.; McCormick, A. V.; Schmidt, L. D.; Cussler, E. L. Ammonia Synthesis Enhanced by Magnesium Chloride Absorption. *AIChE J.* **2015**, *61* (4), 1364.

472974 Production of Acetic Acid from Natural Gas with Zero Carbon Emissions

Abstract:

In this work, the design of a process plant flow sheet that produces acetic acid from natural gas with zero carbon emissions is developed. Two such cases are developed - the production of acetic acid alongside power generation and hydrogen respectively.

The UNISIM software (Honeywell trademark) is used to create the flow sheet for the various modules of each plant based on established technologies of industrial acetic acid and hydrogen production processes. The various modules, when coupled for each process plant, create the net effect of zero-carbon emission.

Energy self-sufficiency, feasibility and operability limits of both cases are established via a thermodynamic analysis, while profitability is ascertained via economic and revenue studies based on current prices of the various commodities. The results are compared to stand-alone power, hydrogen and acetic production plants.

469218 Ionic Liquid Incorporation for Self Healing Purposes: New Measuring Methods for Bitumen Binder, Applications to Asphalt Concrete and Understanding of Chemical Interactions

Abstract:

Asphalt concrete is the most commonly used material in pavement industry. It is prepared by mixing bitumen, which is the bottom product of crude oil distillation, and aggregates at temperatures around 180 â°C. Asphalt concrete, once paved in the roads, experiences traffic loading along with changes in the ambient temperature throughout its service life. Thereby, failures in asphalt pavement such as fatigue cracking, rutting and thermal cracking, occur due to these continuous loading and climatic conditions. The mentioned failures in the roads give rise to building of new roads or maintaining the damaged roads. Building new roads or maintaining the already built ones requires preparation of asphalt mixtures, which consume large amounts of hydrocarbons, bitumen, together with excess energy due to high mixing temperatures. In addition to the large resource and energy consumptions, heating the mixture to high temperatures results in emission of greenhouse gases and bitumen fumes into the air. Therefore, extending the service life of asphalt pavements is very important both from an economic and environmental point of view. Such an improvement would also mean more efficient usage of the fossil resources.

Bitumen, which is the binder of asphalt concretes, is known to have intrinsic self- healing properties. It has the ability to restore its mechanical strength by closing the cracks at high temperatures or during long rest periods. However, during its service life, the asphalt pavement may not always have enough resting periods or the high temperatures which are needed to heal itself. Therefore, improving the self- healing capabilities of bitumen/asphalt is a necessity to minimize the mentioned environmental and economic problems.

In this study, the effects of ionic liquids, which are also known as green solvents, on the self- healing properties of bitumen is studied. 50/70 penetration grade bitumen from Tupras Ä°zmit Refinery is used in the studies. As part of the ionic liquid characterization, Thermogravimetric Analysis (TGA), Differential Scanning Calorimetry and Fourier Transform Infrared Spectroscopy are carried out. The bitumen is characterized by means of TGA, FTIR, DSC, Elemental Analysis and Saturate- Aromatic- Resin- Asphalthene (SARA) fractionation. Two new test methods are developed in order to measure the self- healing capability of bitumen. One of these methods is developed to see the effects of rest periods when intermittent loading is applied on the sample at high temperatures. The other method is developed to investigate the effects of long rest periods on the self- healing capability of bitumen at low temperatures. Measurements are carried out both on unmodified and ionic liquid modified bitumen to investigate the effects of ionic liquids with different side chain lengths on the self- healing properties of the bitumen. In addition to the measurements by the newly-developed techniques, stripping tests are carried out to find out about the effects of different ionic liquids on bitumen- aggregate interactions. Zeta potential measurements are done to quantify the electrostatic interactions between bitumen and aggregates. The asphalt performance of the modified bitumen is measured by Universal Testing Machine with Fatigue test, in which a loading- rest period cycle is applied to the sample until the sample is cracked, and Stiffness tests, which measures the hardness of the asphalt sample.

The results indicated that different ionic liquids have different effects on the bitumen/asphalt properties. With the addition of ionic liquids into bitumen, a 25% improvement in stripping properties, 30% improvement in high temperature self-healing performance, 200% improvement in low temperature performance and 120% improvement in asphalt fatigue life is obtained, meaning that different modification recipes can be used in road pavements depending on the climatic conditions and traffic density.

470079 Environmentally Responsible Fabrication of Efficient Perovskite Solar Cells from Recycled Car Batteries

Abstract:

Environmentally Responsible Fabrication of Efficient Perovskite Solar Cells

from Recycled Car Batteries

Po-Yen Chen[1,2]*, Paula T. Hammond[1], Angela M. Belcher[3]

[1] *Department of Chemical Engineering, Massachusetts Institute of Technology, Cambridge, MA, 02139*

[2] *School of Engineering, Brown University, Providence, RI 02912*

[3] *Department of Materials Science and Engineering, Massachusetts Institute of Technology, Cambridge, MA, 02139*

Email: pychen@mit.edu

Organolead halide perovskite solar cells (PSCs) show great promise as a new large-scale and cost-competitive photovoltaic technology. Power conversion efficiencies over 15% to over 20% have been achieved within 18 to 36 months of development, and thus perovskite materials have attracted great attention in photovoltaic research. However, the manufacture of PSCs raises environmental concerns regarding the over-production of raw lead ore, which has harmful health and ecological effects. In this study, we have demonstrated an environmentally responsible synthetic pathway to reuse car batteries for the fabrication of efficient PSCs (**Figure 1**). Both the anodes and cathodes of car batteries serve as material sources for the synthesis of lead iodide perovskite materials. In contrast to the traditional lead extraction process, our synthesis pathway from recycled battery materials occurs at a lower temperature (600 degree C) and does not include the hazardous emission of lead vapor/dust and CO_2 to the environment. The lead iodide perovskite materials synthesized from car batteries and high-purity reagents demonstrate identical material characteristics. The photovoltaic performance of the PSCs synthesized by each route is the same, which demonstrates that the device quality does not suffer from the materials sourced from spent car batteries. Also, electrochemical impedance spectra (EIS) measurements reveal that each device type displays the same resistances of electron recombination, indicating that the electron-trans- port properties of the lead iodide perovskite are identical. Finally, a simple economic analysis reveals that a single lead- acid car battery can supply enough lead material for the fabrication of 709-m^2 PSCs, which can provide enough electricity to power 30.2 US residential units in Las Vegas, Nevada. With these techniques, the time required to find a lead replacement for PSCs can be further increased. The environmentally responsible fabrication is expected to be broadly applicable not only to the PSC technology but also other applications.

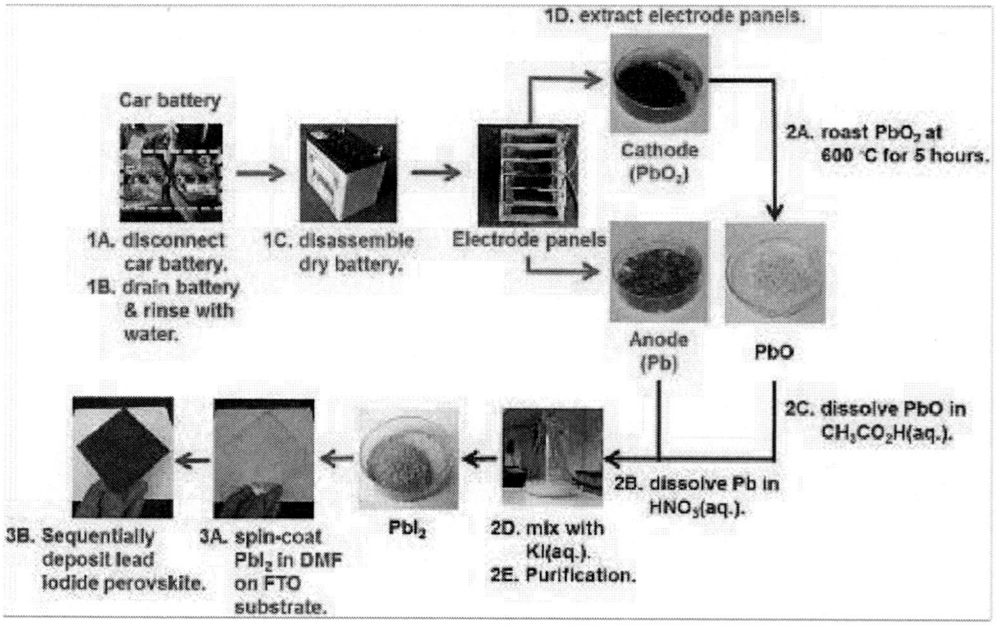

Figure 1. Synthetic process of lead iodide perovskite materials from a lead-acid car battery. The process includes three steps: (1) harvesting material from the anodes and cathodes of a car battery (red); (2) synthesizing PbI_2 from the collected materials (blue); (3) depositing lead iodide perovskite nanocrystals (green).

471578 Recycling & Garbage - a Separations & Materials Challenge

Abstract:

The field of recycling and garbage is largely overlooked by the Chemical Engineering community and has been the realm of mechanical, civil and even electrical engineers. But there is much to learn and apply in terms of Chemical Engineering from this field. Some surprising facts is that 99.9% of the separations at recycling facilities are done by machines - using the densities and shapes of materials to perform the separation. There are also major opportunities to increase the profit in garbage and recycling through better separation techniques and innovative material applications.

San Francisco has the goal of not sending anything to the landfill by 2020 - only 4 years away. Right now, they are currently around a 70% diversion rate. As the deadline approaches, innovations in this field cannot be just limited to the realm of mechanical engineering. For instance, in the past 10 years, composting technology has increased the amount of materials which are able to be composted. First only yard trimmings were accepted, then food scraps, now San Francisco is able to accept food stained paper products. The ultimate goal of composting is â€œanything which was a plant.â€

The garbage stream is also seeing innovations. Earlier this year, San Francisco adopted an organic press, which squeezes all the liquids from the garbage with a hydraulic press. Liquid from the process is sent to a biogas plant to convert into methane & then electricity.

On the recycling side, more materials and separation advances are required to keep the recycling industry afloat.

Currently, the driving force for places where recycling makes sense is areas with not a large amount of space - therefore landfill tipping rates (the cost to dump at a landfill) increase. The Pacific Northwest is home to the highest tipping fee in the nation ($142/ton) and land rich Texas has the lowest at $5/ton. Recycling programs therefore follow the incentives of the state they reside. Diversion from the landfill makes economic sense when instead of paying

$40/ton (Bay Area price), separating and giving away material for free. The $40 gap is saved but no profit is made.

The real profit from recycling comes from one material - aluminum. It accounts for 2% of the weight of the recyclate but 40% of the revenue. And paper and plastic are shipped overseas to be manufactured. Advances in material separations and new applications of recycled material is required to make recycling a truly sustainable business model.

Glass is the major loser of recycling. It is such a heavy material so the cost of transportation and disposal is very high. Additionally, remanufacturing options are limited - making new glass bottles is much cheaper than using recycled glass. New, business-sustainable applications of glass are required to make recycling of glass not a losing proposition. Companies are putting recycled glass into concrete and pavement - which is good - but these are not high margin applications. So, glass needs to be and is heavily subsidized by the State of California - which charges a fee to distributors (and then end customers) for glass products.

These opportunities in the garbage and recycling industries are ripe for the chemical engineering community. The focus must turn towards novel applications of materials and advanced chemical separation techniques â€" both well within the chemical engineering realm.

478273 Characterization of Cationic Poly(N-isopropylacrylamide- co-N- 3-aminopropyl(methacrylamide)-Co- Methacryloyoxybenzophenone) Thin Film Hydrogels

Abstract:

Poly(N-isopropylacrylamide) or PNIPAAm in aqueous solution is known to display a lower critical solution temperature (LCST) undergoing a reversible coil-to- globule transition at 32-33°C. Above this temperature, the polymer switches from being hydrophilic to hydrophobic, and coalesces out of solution. It is well known that small amounts of salts copolymerized with PNIPAAm can raise or lower the LCST of NIPAAm significantly. Moreover, copolymerization of NIPAAm with other monomers can introduce reactivity to stimuli other than temperature, such as pH, ultraviolet (UV) light, magnetism, electric current, etc. In the present work, the influence of a cationic co-monomer, N-3-aminopropyl(methacrylamide) (3AMA), was investigated on the swelling properties of thin film hydrogels of PNIPAAm copolymerized with methacryloyoxybenzophenone. The copolymers showed reactivity to both pH and temperature depending on the percentage of the 3AMA. The cloud point

properties of the bulk solutions were measured and compared to the swelling response of the corresponding hydrogel film. The films were prepared by spin coating the copolymer and then irradiation with UV light was used to induce cross-linking of the benzophenone groups. The swelling behavior of the thin hydrogel films was characterized with both ellipsometry and FTIR.

477679 Solar Thermal Adsorption Refrigeration

ABSTRACT:

Unreliable electrical supply to medical clinics in remote areas of India makes it difficult to safely store vaccines and other medications. The ETHOS (Engineers in Technical Humanitarian Opportunities of Service-learning) program at the University of Dayton, in conjunction with the SAAP (Solar Alternatives and Associated Programmes) group in Bihar India, are developing a novel solar-thermal adsorptive refrigeration system for use at these clinics that does not require electricity and uses safe, environmentally friendly, and widely available materials for the refrigerant and adsorption media. This refrigeration system uses ethanol as a refrigerant and activated carbon as the adsorbent, and can achieve refrigeration temperatures as low as $4°C$. This presentation will describe the technical challenges associated with an ethanol/carbon refrigerator and present the results of recent experimental work on the design and operation of this novel refrigeration system.

477676 Mechanics of Drying in Fluidized Beds

Abstract:

Pharmaceutical drying is an important part of tablet production because of its impact on final product characteristics, both on the API as well as the excipient. The main function of drying is to remove unwanted moisture during tablet formulation as it compromises the mechanical strength and longevity of the tablet. Drying is primarily implemented in the pharmaceutical industry as a batch process, despite the limitations it has compared to a continuous process. Since continuous drying is a relatively new process, there is a lack of full understanding of the heat and mass transfers that govern the process. This lack of understanding makes it difficult to scale-up the process from lab scale to industrial scale.

A fluidized bed is a relatively new technology now commonly used in industry to dry pharmaceutical ingredients. To understand how process operation parameters and heat and mass transfers work in a continuous fluid bed, it is important to learn their effects in a batch process. Experiments were carried out by wetting dibasic calcium phosphate anhydrous (DCPA) with distilled water to varying moisture levels inside a fluidized bed using top spray. Once the desired moisture content was reached, the wet powders would be dried in the fluid bed at 80°C. The moisture content would then be recorded at 3 minute intervals during drying. The results showed that there were three stages to the drying process: an initial preheating period, a constant period, and a falling rate period. Experiments were also done to see the effect of particle size on the drying rate. The results showed that there was a negligible difference in the drying rate when the particle size changed, contradictory to what previous studies have shown. Thus, more experiments are needed to understand this behavior.

Further work will be done on the continuous fluid bed as well as making sure the results obtained from the batch fluid bed are precise and accurate. The first study to be carried out on the continuous fluid bed, which behaves as a semi-plug flow reactor will be a residence time distribution. This will be useful in characterizing the fluid bed, as well estimating the yield of future studies and designing future fluidized beds. The data obtained from these studies will help make continuous fluid beds implementable on an industrial scale and thus, will greatly reduce the energy and time required to manufacture tablets as well as lower the consumer price of costly medicine.

477674 Engineer Microogranism Escherichia coli to Produce Biopolymer Precursor Cadaverine from Renewable Resources

ABSTRACT:

The global market demands a huge amount of polyamide, most of which is produced by the petroleum industry. However, new approaches for high-efficiency production of polyamide is in a pressing need due to the economical, environmental and sustainability concerns for reliance on petroleum resources. This research aims to develop a microbial synthesis method to produce an important biopolymer precursor, as an alternative approach for the conventional petrochemical method for producing polymers. Specifically, we metabolically engineered *Escherichia coli* to produce biopolymer precursor, cadaverine (1, 5-diaminopentane) from simple sugar, which can be derived from biomass or other renewable feedstocks. A *ldcC* gene encoding a lysine decarboxylase was first cloned in a plasmid using genetic engineering methods. The resulting plasmid was subsequently introduced into *E.coli*, resulting in an engineered strain that can convert simple amino acid lysine to cadaverine. The integration of such cadaverine biosynthesis with lysine formation from simple sugars offer a complete biosynthetic route for cadaverine production from renewable sugars. As there have been extensive efforts in industry for exploitation of bio-based green materials, it is anticipated that our research will contribute to the generation of more applicable and robust techniques for manufacturing of non-petroleum based products.

Bacterial Biofouling Prevention by Magnetic Particles/Polycardanol Composite

John Hoffman

Senior Undergraduate

Department of Chemical and Environmental Engineering

The University of Toledo

COLLEGE OF ENGINEERING
THE UNIVERSITY OF TOLEDO

Outline

- Introduction and Theory
- Method and Procedure
- Results
- Conclusions
- Future Work

THE UNIVERSITY OF TOLEDO 1872

Introduction- Biofouling

- Biofouling is an area of concern that affects many industries, most notably the chemical, marine, and medical fields.

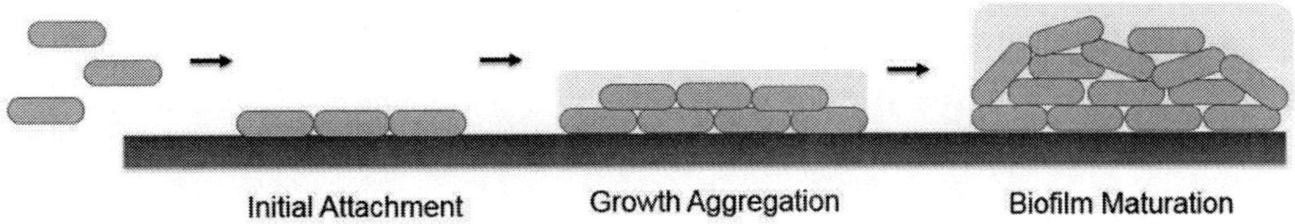

THE UNIVERSITY OF TOLEDO 1872

Introduction- Preventing Biofouling

- There are three methods for preventing biofouling:
  - Surface Modifications
  - Contact Killing
  - Biocide Leaching
- Our research attempted surface modifications as it doesn't have adverse affects on the surrounding environment.

THE UNIVERSITY OF TOLEDO 1872

Introduction- Polycardanol

- Polycardanol is an enzymatically polymerized compound of cardanol, a phenolic monomer derived from cashew nut shell liquid and has shown in previous research to be antibacterial and anti-fouling.
- We have expanded on this knowledge to create polycardanol coatings with ferrite magnetic particles of various sizes to determine if this increases the anti-fouling nature of polycardanol.
- Fluoropolymer was added to the polycardanol coating to test if increased surface hydrophobicity decreases bacterial attachment.
- PEG was also evaluated after reviewing the fluoropolymer results to test a surface with decreased hydrophobicity.
- These coatings were tested in a flow cell over a three hour time period to understand how the coatings affect bacteria attachment in flowing conditions.

THE UNIVERSITY OF TOLEDO 1872

Method

- First a bacteria solution is created.
  - *Pseudomonas aeruginosa* is placed into a nutrient broth and let to incubate for 18-24 hours before the experiment.
- The bacteria is then centrifuged and suspended in a phosphate buffer solution, with an optical density of 0.05.
  - This correlates to a bacteria solution with 5-20x10^9 cfu/mL bacteria this is verified by serial dilution and plate counting.

THE UNIVERSITY OF TOLEDO 1872

Method

- The bacteria solution is then run through the flow cell over the coated slides for a period of three hours at 1mL/min.
- Images are taking of the slides to count to number of bacteria attachment.
- Bacteria is manually counted and normalized to glass bacteria attachment.

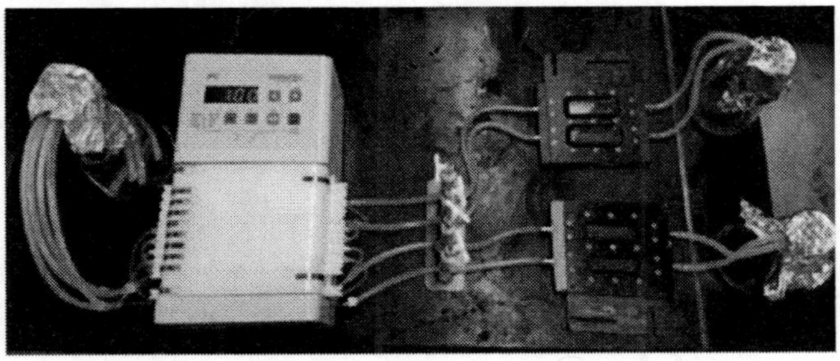

THE UNIVERSITY OF TOLEDO 1872

Results- Magnetic Coating

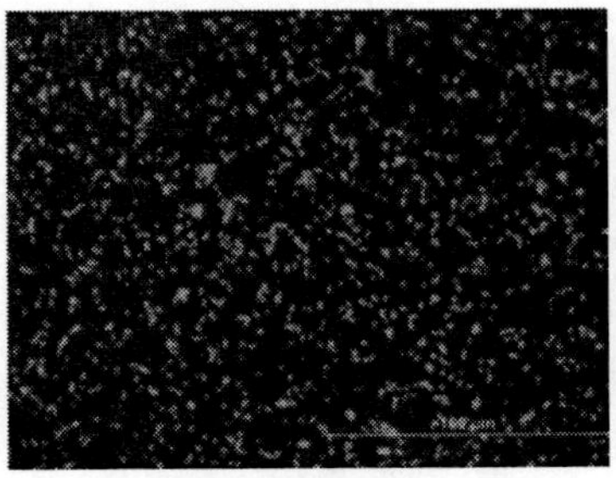

Glass after 3 hours

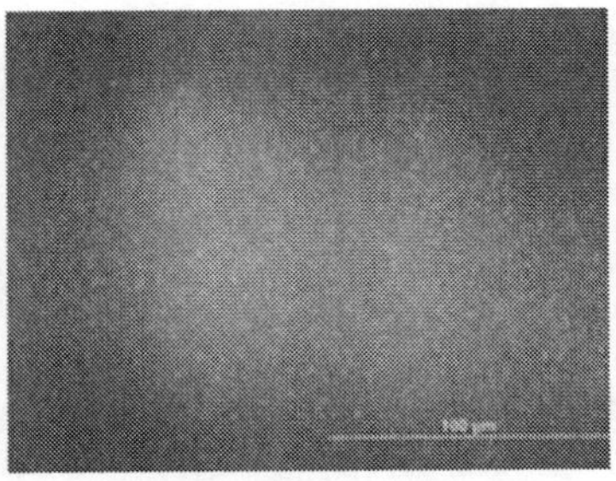

Polycardanol after 3 hours

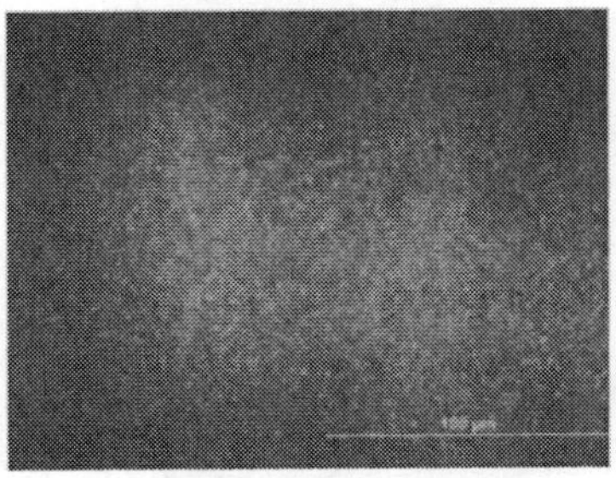

Lanthium Strontium Ferrite after 3 hours

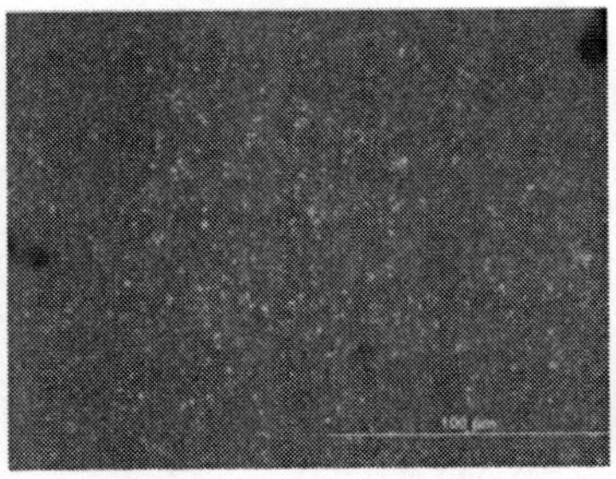

Cobalt Iron Oxide after 3 hours

THE UNIVERSITY OF TOLEDO 1872

Results- Magnetic Coating

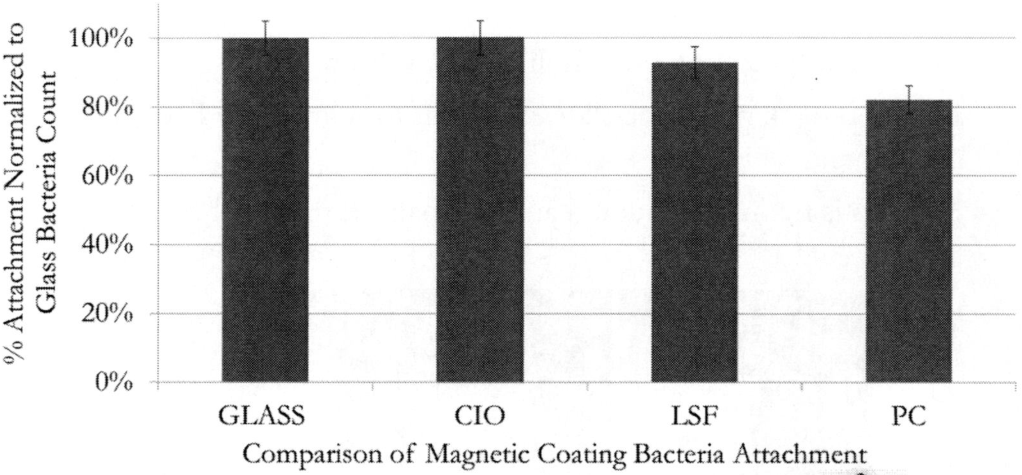

Results- Fluoropolymer Coating

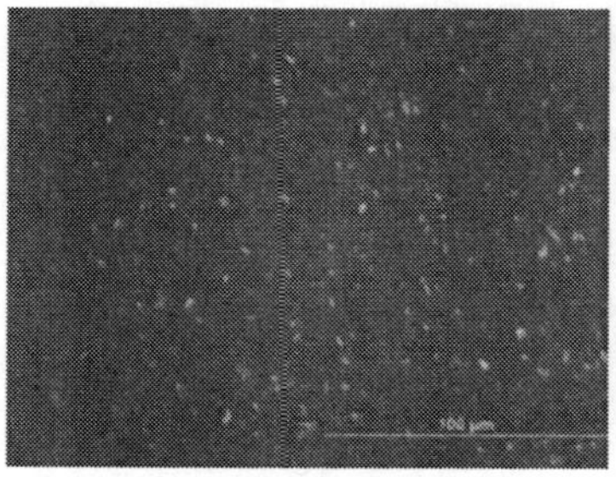

0.5% Fluoropolymer after 3 hours

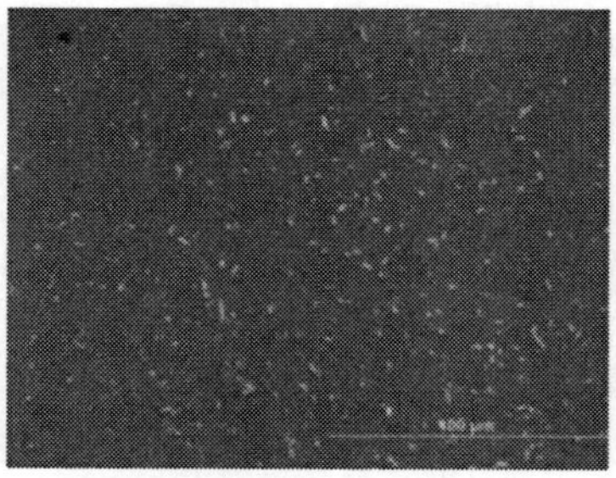

0.1% Fluoropolymer after 3 hours

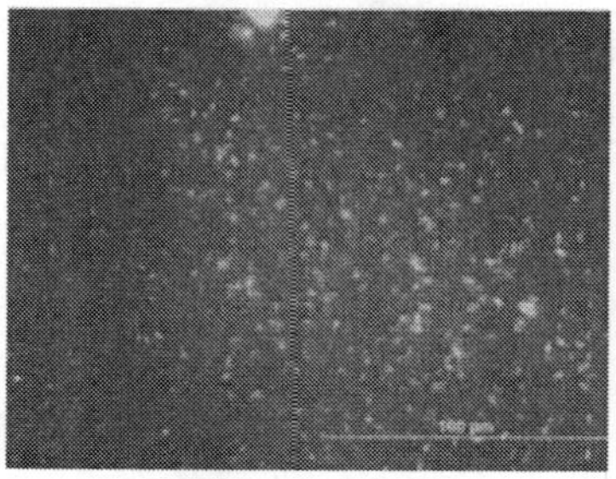

2.0% Fluoropolymer after 3 hours

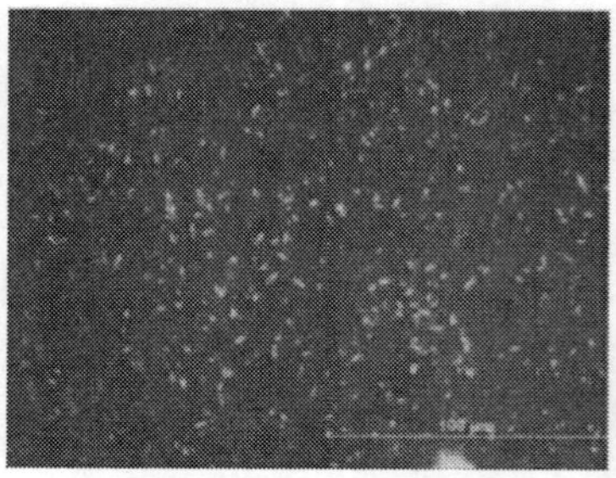

5.0% Fluoropolymer after 3 hours

Results- Fluoropolymer Coating

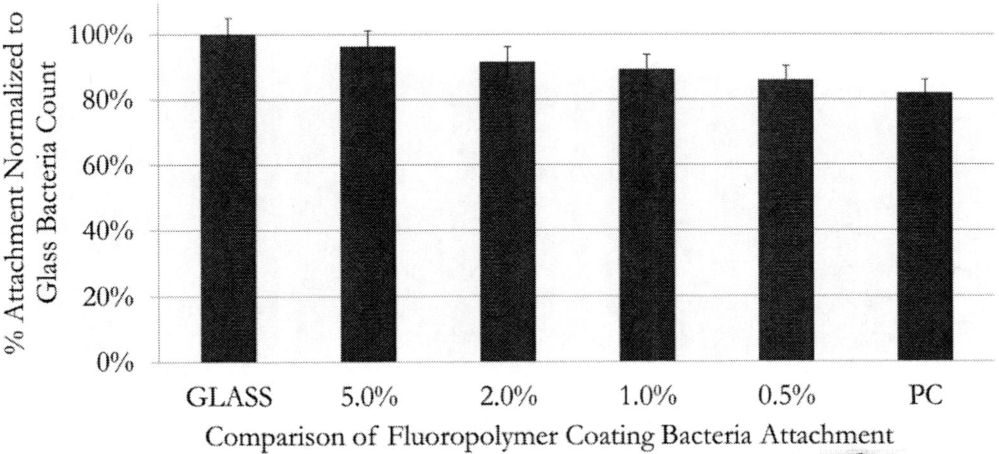

Results- PEG Coating

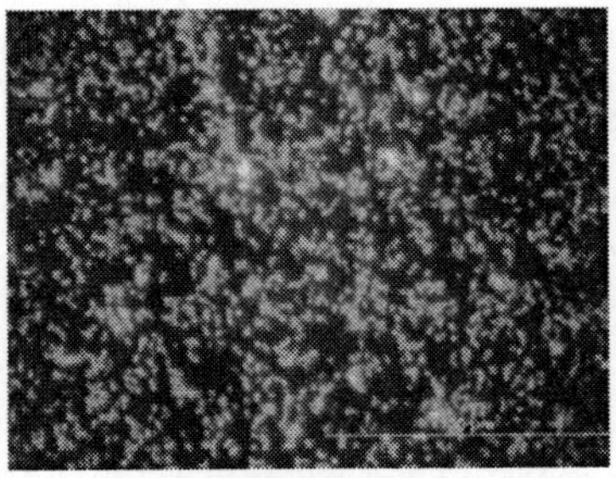

Glass after 3 hours

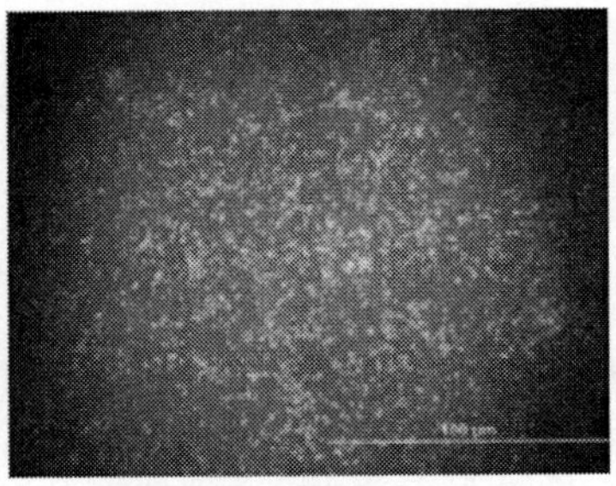

Polycardanol after 3 hours

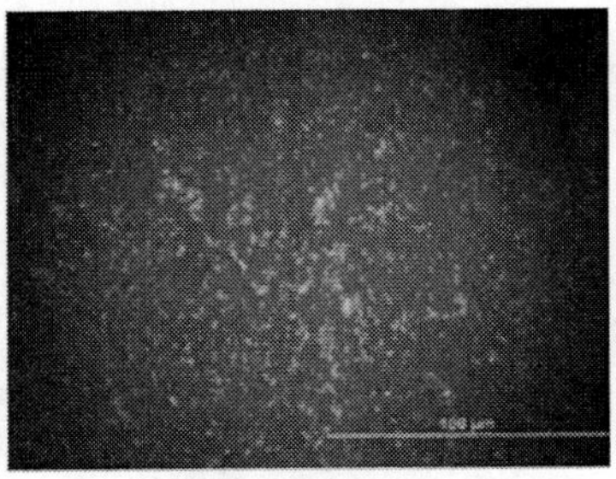

5% PEG after 3 hours

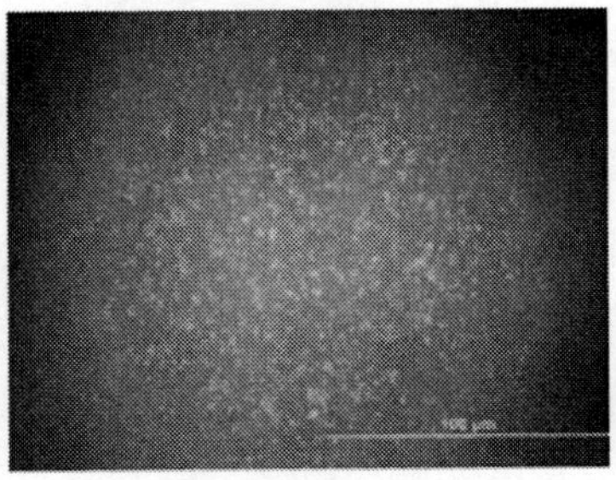

10% PEG after 3 hours

Results- PEG Coating

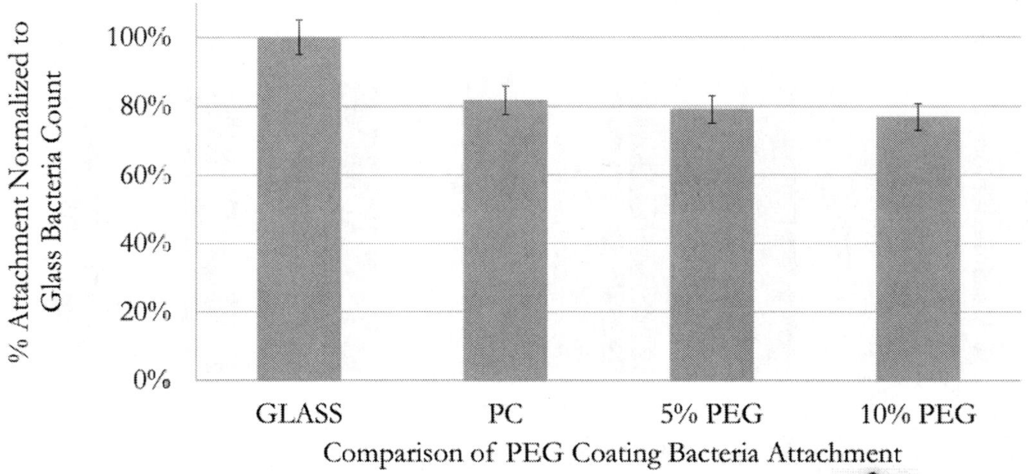

THE UNIVERSITY OF TOLEDO 1872

Conclusions

- Magnetic particles did not express the believed antifouling characteristics.
  - Can be correlated to increased roughness of magnetic coated slides.
  - Microscopic surface topography is a key contributor to surface attachment.
- Addition of the fluoropolymer compound to the coating had the opposite affect towards biofouling.
  - Demonstrates that increased hydrophobicity did not reduce bacterial attachment.
- Addition of PEG to the coating confirms that decreasing hydrophobicity results in reduced bacterial attachment.

THE UNIVERSITY OF TOLEDO 1872

Future Work / Recommendations

- Develop method for testing magnetic particles with similar surface characteristics to polycardanol.
  - Magnetic particles suspended in a polymer matrix would reduce surface macrostructures.
  - Develop better testing method for magnetic force, results showed negligible force.
- Test coatings with an implant grade plastic as a control instead of glass, with comparisons against a current industry anti-fouling coating.

THE UNIVERSITY OF TOLEDO 1872

Acknowledgements

- Dr. Dong-Shik Kim
- Lohit Sharma
- The University of Toledo Chemical Engineering Department
- OUR-UT Undergraduate Research USRCAP Program
- Ohio Third Frontier Program

References

- Boonsai, P., Phuwapraisirisan, P., Chanchao, P. (2014) . Antibacterial activity of a cardanol from Thai Apis mellifera propolis. International Journal of Medical Sciences 11(4):327-336.
- Chelikani, Rahul. (2007) Development of Novel Antibiofouling Materials from Natural Phenol Compounds. Proquest

THE UNIVERSITY OF TOLEDO 1872

Questions?

THE UNIVERSITY OF TOLEDO 1872

477673 Functionalized Liposome Delivery Targeting Opportunistic Fungi

Abstract: Given the steady increase of drug resistance, there is a need for targeted drug delivery systems that limit unnecessary exposure to antimicrobials. This project explores the development of a novel liposomal formulations targeting *Candida albicans*, an opportunistic fungal pathogen. *C. albicans*, the most common cause of fungal bloodstream infections,[1] can give rise to systemic candidiasis, an infection with a 33-54% mortality rate.[2] To specifically target these fungal cells, a peptide fragment of histatin 5, which binds to the *C. albicans* cell wall,[3] will be incorporated into liposomes. These functionalized liposomes should minimize damage to healthy cells, limit antimicrobial exposure and reduce the development of antifungal resistance.

The modified peptide fragment of histatin 5, AQRHHGYKRQFH, was formed using solid phase peptide synthesis with standard Fmoc chemistry. Palmitic acid, a saturated fatty acid, was then conjugated to the N-terminus of the peptide sequence. Liquid chromatography and mass spectrometry (LC-MS) confirmed the successful synthesis of the peptide-palmitic acid conjugate (See Figure 1). This peptide-palmitic acid conjugate was then combined with two clinically-used lipids, 2-dimyristoyl-sn-glycero-3-phosphocholine (DMPC) and 1,2-dimyristoyl-sn-glycero-3-phosphorylglycerol (DMPG) to form liposomes through a freeze-thaw cycle and extrusion process. Dynamic light scattering confirmed the formation of liposomes with a uniform size of 141.5 ± 0.6 nm diameter and a polydispersity index of 0.102 ± 0.013 (See Figure 1). In conclusion, the development of a functionalized liposome has been confirmed. Further work will incorporate anidulafungin, a hydrophobic drug currently administered intravenously without a carrier. Tests will be performed to determine the release profile, the antimicrobial efficacy and the stability of these liposomal formulations.

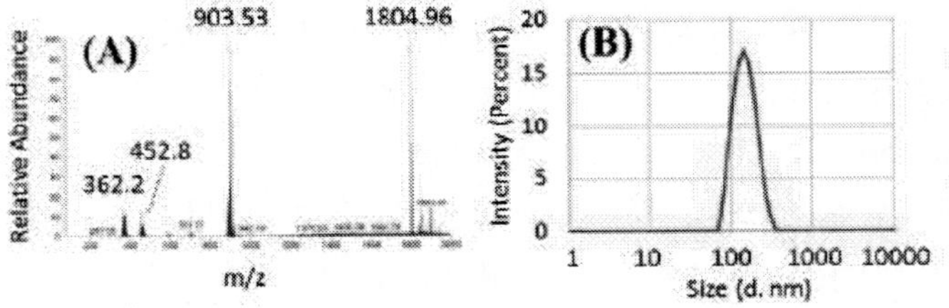

Figure 1. (A) Mass spectrometry of the palmitic acid-peptide conjugate shows prominent peaks at 1804.96, 903.53, 452.79, 362.21 m/z. These represent the 1+, 2+, 4+ and 5+ ionized states of the peptide sequence. (B) DLS confirms the formation of the liposomes with incorporated peptide with a single peak at 141.5 ± 0.6 nm diameter and a PDI of 0.102 ± 0.013.

477670 Evaluating the Effect of Triacetin on the Cold Flow Properties of Fatty Acid Methyl Ester Blends

Abstract:

Biodiesel consists of mono alkyl esters of long chain fatty acids derived from a renewable lipid feedstock, for use in compression ignition (diesel) engines. Biodiesel offers advantages including low toxicity, lubricity, optimal flash point, low sulfur content, lower exhaust content, and biodegradability. While there are several advantages to using biodiesel, some potential technical hindrances are associated with the fuel properties, including its functionality in cold weather climates.

Optimizing the cold flow properties of biodiesel is therefore necessary for its application as an alternative to conventional petrofuels. This research examines the cold flow properties of binary fatty acid methyl ester (FAME) blends both with and without triacetin, a potential cold flow improver. The cloud point, melting point, and heat of crystallization of various blends of FAME and triacetin were analyzed to examine crystallization phenomena. Generally, the cloud point of pure FAME decreased with decreasing carbon number and increasing degree of unsaturation. Furthermore, the incorporation of triacetin as a cloud point depressant was studied in both pure component and binary FAME fuel blends. When introduced in a weight proportion of up to 20 wt%, triacetin was observed to depress cloud point by up to 2.7 K. Thermodynamic analysis and predictive modeling demonstrated that cocrystallization affects the cloud point of binary FAME blends. Based on the observed results, triacetin may depress the cloud point of a system because of its role as a diluent rather than as a crystal modifier. It is concluded that incorporating triacetin in FAME or biodiesel blends can moderately improve cold flow behavior.

477669 Solvents Effects on the Structure of Asphaltene Aggregates

Abstract:

Controlling asphaltene deposition is a critically important topic in petroleum production, but asphaltene aggregate behavior remains poorly understood even with an abundance of hypotheses and speculations concerning this topic. Current research suggests classifying asphaltenes into 2 types: soluble and insoluble based on their fractal dimension when precipitated in the presence of heptane. Under these conditions two general fractal dimensions have been observed: ~2.1 for insoluble asphaltenes (those that grow to the micrometer scale and precipitate), and ~1.7 for soluble asphaltenes (those that remain on the nanometer scale and stay suspended in solution)1

Rheological measurements made in the lab also show that the intrinsic viscosity of asphaltene aggregates changes as a function of heptane concentration. This observation is in agreement with prior research that examined the molecular weights and intrinsic viscosities of asphaltene aggregates in the presence of heptane2

. Indeed aggregate structure is a function of the solvent
environment, and this structure determines the phase behavior which is a precursor to deposition. This talk will present research showing the changes in intrinsic viscosity and fractal dimension as a function of flocculent concentration, and address the competing causes including aggregation and solvation effects. A better understanding of aggregate behavior on the nanoscale is critical to developing a more inclusive predicative model. Improving the understanding of asphaltene behavior will ultimately benefit the petroleum industry by helping to avoid costly shutdowns due to clogging and equipment failure.

R&D Management from Beginning to End

Jack Hipple
Principal
Innovation-TRIZ
Tampa, FL

Eldon Larsen
Professor of Engineering
Marshall University
South Charleston, WV

Extended Abstract

This workshop focuses upon managing R&D projects from beginning to end. Beginning refers to mental picture of an idea whereas end refers to solving the customers need and in the process generating profit for the organization. Using a case study, the speakers will delineate challenges involved at every stage of R&D project from a mental image of an idea to a profit making product. The talk also compares old paradigm of innovation management with the new proposed paradigm. The probability of success can be significantly improved using the new paradigm.

Successful Innovation Management

We have seen many idea management flow sheets, gate management processes, and prioritization recommendations starting with an idea that already exists, but no discussion of where the idea came from or what was necessary to generate and frame it. Moreover attention is not paid to shaping and propagation of the idea. This has been especially true over the past decade when processes such as Six Sigma came into vogue whose main thrust was to eliminate deviation. New concepts and innovation are, by their very nature, deviant! In order to generate fresh and innovative ideas we must create a new thought pattern that may challenge the existing status quo in any number of ways. These can even include, at an extreme, replacing one's current products and businesses with totally new approaches and models.

The key elements of successful commercial innovation include:

1. Continuous flow of unmet needs, both from current customers and related to core competencies of the organization ("outside in" as well as "inside out"
2. Using clear mental models of problems and potential solutions
3. Perseverance for idea propagation
4. Clear definition of the business opportunity

5. An appropriate product development and management team
6. Efficient process scale up and continued input from the market
7. Successful project management to insure implementation on schedule and within budget

We will look at each one and discuss the important issues.

1. Continuous Flow of Unmet Needs and Exploration of Uses for Core Competencies

An organization needs to have a process for how it decides what to work on. This may sound simplistic, when this is left to intuition or accident; it is unlikely that the "best" projects will be chosen.

Steps to make sure that an innovation effort is capturing the best opportunities:

- Talk to groups that are not your current customers but that might be able to use the functional capability that you have as a core competence. Customer competitors are another valuable source. The customers of your customers are another excellent source. This insures that you have awareness several levels above your product and customers and alerts you to potential long term needs and trends that may not be obvious to your direct customer or may not be readily disclosed by them.

In these discussions we want to make sure that we have a clear definition of what the "ideal result" that our customers are desiring. We can also look at parallel universes of technology and markets to explore opportunities.

2. Using Clear Mental Models of Problems and Solutions

It is one thing to "invent" a new product, but another to make sure that we understand how a potential customer is going to use the product as well as how the customer's customer is going to use their product. The days are gone when we can just make something, put it in a drum or tank car, and wait for the invoice to be paid. We must be intimately familiar with the product's use, its limitations (including regulatory issues if appropriate). We must understand the details of the physical use of the product and how it affects a customer's products and business.

It is also important to make sure that we do not have blinders on when thinking about use of our product. It is rare that there is only one way to accomplish a function. We must be constantly vigilant about competitive activities, not just from direct competitors, but from parallel approaches to meeting a customer's needs.

3. The People Side

Ideas, innovation, and projects never happen in a vacuum. They are accomplished with teams of people, all of which have different personalities and styles. Recognizing and using these pro-actively is critical to a project's success. How these are measure and used will be discussed.

4. The Business Opportunity

A new product is not a business unless we have a way of sustaining the product's sales, expanding its utility, and developing and enforcing an intellectual property portfolio. As we commercialize a new product, we must be constantly aware of, and be willing to address with resources and commitment, issues such as change in product form, packaging and shipping issues, product purity, and shelf life/stability. These issues are most adequately addressed by constant and continuous communication with the customer.

5. Product Development Teams

We frequently choose product team members based on "who's available". While this may be expedient, it may not serve the best interests of the new business development. The obvious component of team selection is function—marketing R&D, manufacturing, legal, sales, etc. but we frequently ignore the human side of team functioning. If team members spend too much time dealing with personal issues relating to other team members, that energy is not applied to successful new product launch. Tools that can assist in this include measurements such as the Myers Briggs family of instruments, the Hermann Brain Dominance instrument (HBDI), and the Kirton KAI assessment. Knowledge of these personality profiles will not change natural behavior, but can make team members aware of others' preferred styles and open discussion can eliminate hidden barriers to effective team functioning.

6. Efficient process scale up and continued input from the market

Often times, companies strive for launching the best product. It is important to realize that continued input from the market is critical in order to have that 'best product' in the market. It is necessary to efficiently scale up and launch a product sooner than later. It should be perfected using market feedback. Getting market feedback is the only way to close the R&D loop and serve your customers to the fullest.

7. Successful Project Implementation and Management

New product development does not end when the first semi-commercial sample has been OK'd by the customer. A new type of team is now required to oversee the details process design, plant construction (including all the sub issues of permitting, EPA regulations, etc.). Quick and efficient plant start up with appropriate contingency planning is also a part of this effort. Contingency planning, not only related to unforeseen plant construction and start up difficulties, but also to taking care of the customer's needs in these situations must be part of the plan.

Three keys to successful project management lead to successful project completion. These keys involve

- The importance of people
- The importance of defining and setting good objectives
- The importance of good planning

We will apply these three keys to successful project management. This is all about application, not just theory.

475804 A Green Chemistry Approach for Producing Non-Synthetic Pesticide in Under-Developed Regions

Abstract:

The purpose of this contribution is to present the results of a US EPA funded People, Prosperity and the Planet Design Competition Project to develop an appropriate green chemistry based, low-cost method for producing a non-synthetic pesticide, wood vinegar, from already existing biochar production techniques present in rural regions of in Cameroon, Uganda and India. The overall cost to construct the wood vinegar processing unit must be low (typically less than 100 USD), construction materials must be readily available and construction methods must be simple and easily explainable to people with little or no formal education. The purpose behind this economic constraint is to give accessibility to even the poorest of the target population.

This project is based on applying the principles of sustainability, appropriate technology and green chemistry with chemical and mechanical engineering to develop a technology and processes that will provide rural people in developing countries with the ability to produce wood vinegar in addition to their production of biochar. To achieve this objective, four tasks were completed:

- Design and optimize a production method for generating wood vinegar by pyrolysis of locally available biomass.
- Determine the composition of the wood vinegar and its variance as different biomasses are pyrolyzed.
- Conduct field studies in Cameroon, Uganda and India for the use of wood vinegar as a pesticide, herbicide, and fungicide to enhance its effectiveness.
- Mechanical design and fabrication plans for making wood vinegar using locally sourced construction materials to implement the prototype together with our research partners.

Because the individuals in Cameroon, Uganda and India were already familiar with the process of making biochar, this research will be especially important since the process proposed by the research team adds a value added co-product to the production of a product that already enjoys widespread use. This is a key hallmark of sustainable product design.

This project is a co-recipient of the 2016 AIChE YCOSST P3 Award.

476362 Driving the Future of US Biotechnology: Governance and Ownership Challenges in the Age of Gene Drive

Abstract:

More than 3 billion people live in regions afflicted by malaria and other mosquito-borne diseases. Despite tremendous efforts, the World Health Organization estimates between 655,000 and 1.2 million people die from malaria each year, many of them children. Traditional vector control mechanisms, such as Sterile insect techniques or habitat removal, suffer from high economic and human costs. The recent development of CRISPR/Cas9 gene drives offers a powerful new means for controlling the spread of vector borne disease by propagating traits that interrupt the vectorâ€™s ability to serve as a host for disease. Gene drives work by aggressively converting organisms heterozygous for the drive and associated traits into ones homozygous for the drive and associated traits. This causes a target allele to spread rapidly throughout a population. Given their propensity for self-propagation, these systems also radically challenge the existing governance structures for biotechnology.

In this talk, I outline recommendations for governing this emerging technology, focusing specifically on issues relating to oversight and liability, confinement and monitoring, and biosecurity and risk assessment. Gene drive challenges our existing technology paradigms; as such, a holistic approach is needed that centers on public engagement at the local, national, and international level. Similarly, holistic approaches are needed in researching and governing this technology: funding agencies must not only focus on the engineering questions, but on the ecological ones as well. Regulators should expand their scope to include additional federal agencies underneath the Coordinated Framework for the Regulation of Biotechnology, and risk assessors should operate in tandem with threat assessors to understand the opportunities presented by this technology in the context of relevant challenges. Gene drives have significant potential to address major problems in public health by preventing the ability of mosquitos to serve as vectors for disease. However, this technology is still in its infancy and much must be done to engage the public and develop appropriate controls and governance regimes prior to proposed environmental release.

476364 An Aging Infrastructure: Policy Recommendations to Modernize Aging Water Distribution Networks to Protect Human Health

Abstract:

The water distribution infrastructure in the United States is degrading as its lead-based piping approaches its 100-year lifespan and therefore poses substantial health risks to the consumer. With that, regulations are not adequately designed to keep pace with system-wide corrosion and do not facilitate proper infrastructure investment of pipes in-service; an example of this is the Flint, MI, water crisis. However, some cities have met their institutional responsibilities. Madison, WI, is such an example of a city able to meet its regulatory and consumer obligations. Juxtaposing these two examples, a clear set of policy recommendations can be developed. Moving forward, it should be the role of the federal government to incentivize reinvestment into water distribution infrastructure in critical condition. Crafting regulations that are designed to promote regulatory synergy from the local, county, to state levels will prioritize this revitalization of our water supply and ensure this cannot happen again.

EMBRACING OUR BIOECONOMIC FUTURE: ENGENDERING POLITICAL AND ECONOMIC SUSTAINABILITY FOR THE ENERGY AND CHEMICAL SECTOR OF THE BIOECONOMY

Tomas W. Green, University of Kansas

Executive Summary

The world is being faced with daunting challenges: food, energy, and water are growing more scarce, the climate is changing, and the cheap natural resources that changed the world are now responsible for a threat to life and environmental health. Alternatives to petroleum products and fossil fuels have existed for over a century, but they have always lacked political and economic sustainability. With growing concern and urgency, global communities are attempting to change their respective economies to sustainable alternatives.

The bioeconomy is a vision for a future where biology can be leveraged to meet the needs of a growing population. There are veteran technologies, such as biofuels, and the burgeoning science of employing microbial life to produce carbon-sequestering fuels and chemicals from waste material. The potential of the bioeconomy is promising, especially for American industry. While the technology is maturing and building viability, it can be advantageous to exploit more profitable sectors of the bioeconomy to support it as a whole. Biofuels and biochemicals are interdependent- they share many of the same challenges and opportunities. Growing support for one can implicitly help the other, and building economic viability will create political sustainability. However, prominent policy debates present political challenges that, coupled with economic uncertainties, are increasingly difficult to navigate. Serious concern over existing policy, the current state of the automotive industry and fuel distribution system, capricious oil markets, immature bio-based technology, and competing environmental policy priorities all make the conversion to bio-based technologies difficult.

In order to expedite the transition to a bioeconomy in a way that is politically and economically sustainable, there should be strategic prioritization of action and investment by federal agencies, and new policies need to be in place that encourage growth in capable technologies by leveraging the free market. Federal agencies should focus their efforts on bioproducts that can create biofuel profitability, investments that will be mutually beneficial to a variety of technologies in the bioeconomy, capitalizing on regional specializations, and developing consensus on prominent debates. New policy should reform the Renewable Fuel Standard and the Corporate Average Fuel Economy Standard, and craft new policy that encourage market adoption of low-carbon biochemicals and bioproducts. Additionally, the potential of pricing carbon in an economically-minded way should also be taken into consideration.

Relevant Legislation and Initiatives

There are a few bills and initiatives that are most relevant to the current political climate and which will shape most of the dialogue moving forward.

The Renewable Fuel Standard

The Renewable Fuel Standard (RFS) was originally established in 2005 under the Energy Policy Act (EPAct) (US Department of Energy n.d.), and was later expanded in 2007 under the Energy Independence and Security Act (EISA) (US Environmental Protection Agency n.d.). These policies are sometimes referred to as "RFS1" and "RFS2," respectively. Under this policy, renewable fuels are defined as "cellulosic biofuels," "advanced biofuels," "biomass-based diesel," and "total renewable fuel" (Program Overview for Renewable Fuel Standard Program 2015). Fuels are judged based on their respective ability to reduce carbon emissions. To determine this, the Environmental Protection Agency (EPA) conducts a lifecycle analysis that examines the entire carbon footprint of a obtaining, manufacturing, transporting, and using a fuel.[i] Compared to the carbon footprint for petroleum fuels in 2005, cellulosic biofuels must achieve a 60% green house gas (GHG) reduction, advanced biofuels and biomass-based diesel much achieve a 50% GHG reduction, and renewable fuels not under those categories (other biofuels) must meet a 20% reduction in GHGs. Cellulosic biofuel, cellulosic diesel, and biomass-based diesel all qualify as advanced biofuels. Total renewable fuels are all advanced biofuels plus renewable fuels, which for the United States is predominantly corn ethanol (US DOE Office of Energy Efficiency and Renewable Energy 2016).

The greatest accomplishments of the RFS have been the historic decline of petroleum to below 95% of the US transportation fuel portfolio (Martin 2016), large and stabilizing growth of ethanol as a fuel additive and octane enhancer, and the development of cellulosic fuels that have substantially less GHG emissions associated with them. However, there are many reasons why lawmakers feed discouraged since the passing of RFS1 and RFS2. In the Annual Energy Outlook (AEO) of 2016, conducted by the Energy Information Administration (EIA), the projected shortfall is over 19 billion RINs in 2022 (Energy Information Administration 2016), compared to the original congressional mandate. Virtually all the shortfall projected lays in the category of advanced biofuels, which includes cellulosic biofuels (Energy Information Administration 2016).

Low Carbon Fuel Standard

California passed a complementary policy to the RFS called the Low Carbon Fuel Standard (LCFS) at the start of 2007 and was enacted in 2009 (California EPA Air Resources Board n.d.). The LCFS sets carbon intensity requirements for transportation fuels for each year, based on a full life cycle analysis (an analysis of every carbon input in the pathway of fuel production, transportation, refinement, etc.) of a fuel being produced (California EPA Air Resources Board n.d.). The ultimate goal of the policy is to reduce the carbon intensity of fuels by 10% in 2020 (California EPA Air Resources Board n.d.).

The LCFS works by setting carbon intensity requirements that fuel (both petroleum and biofuel) producers need to meet. Obligated parties must deliver quarterly and annual reports to demonstrate compliance, and if a fuel that is produced by an obligated party exceeds the carbon intensity requirement, they need to purchase "credits" to offset their compliance deficit (California EPA Air Resources Board n.d.). Parties who produce alternative fuels (e.g. hydrogen and biogas-derived biomethane) can "opt-in" to sell the credits they generate on the market, as well as natural gas producers (California EPA Air Resources Board n.d.).

Federal R&D

Federal research, development, and demonstration (RD&D) dollars have made a profound impact on the maturity of the biofuel and biochemical industry. In aggregate, the DOE has invested more than $3 billion in programs that advance biofuels, develop feedstocks, convert municipal wastes into resources, produce biopower, and generate a variety of bio-based products (Bioenergy Technologies Office 2016). Some of the most notable investments, as can be seen in the figure above, were in the 2008 Farm Bill, the 2009 Stimulus Bill, and the 2014 Farm Bill. The Food, Conservation, and Energy Act of 2008 (also known as the 2008 Farm Bill) included key investments in biofuels with a strong emphasis on cellulosic fuel technologies (Bioenergy Technologies Office 2016). Importantly, the 2008 Farm Bill strengthened the USDA's BioPreferred Program, which was established in the 2002 Farm Bill (Golden 2015).

---

[i] All lifecycle assessments are based on many assumptions and are therefore imperfect. There is considerable debate over the manner in which the EPA conducts its life cycle analysis, and whether their assumptions are accurate. This will be discussed later, in the section "Current and Prospective Political Climate Surrounding RFS."

The American Recovery and Reinvestment Act of 2009 – the stimulus package that sought to curb the recession – provided funding for biorefineries and other clean energy programs (Office of Energy Efficiency and Renewable Energy n.d.). Specifically, it gave $600 million for integrated biorefineries to demonstrate and deploy new technologies and more than $70 million for research and development in advanced biofuels, including algal fuels (Office of Energy Efficiency and Renewable Energy n.d.). The Agricultural Act of 2014, known as the 2014 Farm Bill, supported a variety of investments in the bioeconomy. Title IX: Energy sought to encourage renewable energy, opportunities in bio-based manufacturing, and to grow the biofuel industry.

An Overview of the Bioeconomy

The term "bioeconomy" was first coined by two geneticists in 1997, who defined it as an "economic field which uses novel biological knowledge for commercial and industrial purposes" (Westra, Gray and D'Aloia 2016). It can be thought of as the "economization of all living entities." More formally, the bioeconomy has been defined as "the knowledge-based production and utilization of biological resources to provide products, processes and services in all sectors of trade and industry within the framework of a sustainable economic system" (German Bioeconomy Council 2015). The size of the bioeconomy is estimated to approximately $350 billion annually (Golden 2015). U.S. revenues from industrial biotechnology (fuels, enzymes, and materials) surpassed $125 billion in 2012 with 1.5 million jobs across all sectors of the supply chain (Golden 2015). Additionally, for every bio-based products job, 1.64 more jobs are created in the US (Golden 2015).

As of 2014, the usage of alternative energy has reached a historic height, with biomass energy serving as the most prolific source of renewable energy (US Department of Energy 2016). Comparative amounts of renewable energy use over time is shown in Figure 4. There is much to be said for the power of biology, and its ability to serve as a source of power and products that will decrease dependence on foreign oil, aid in the fight against climate change, and grow the economy.

The Potential Future for the Bioeconomy

Biofuels have experienced encouraging growth that is expected to continue. Global production has grown 625% from 2000 to 2010 (Kovarik 2013). Currently, biofuels account for 3% of total world road transport fuel, but that is expected to grow to 27% by 2050, cutting down on 2.1 gigatonnes of CO_2 per year (Kovarik 2013). Biomass is expected to be the predominant feedstock of the future of chemicals and transportation fuels (Lamers, et al. 2016). The successes so far are building the potential to further grow the bioeconomy by an additional $100 billion by 2030, with the creation of an additional one million jobs, particularly in rural areas where feedstocks are grown (Subcommittee for Advanced Manufacturing 2016). The National Bioeconomy Blueprint stated that there is "a set of emerging trends in recent research that foreshadow major advances in the areas of health, biological-based energy production, agriculture, biomanufacturing, and environmental clean-up" (Office of Science and Technology Policy 2012).

The U.S. DOE, in collaboration with Oak Ridge National Laboratory, produced a report in 2005 that sought to examine the possibility and scope of a billion-ton bioeconomy as means of replacing petroleum use. The report, entitled "Biomass as a Feedstock for a Bioenergy and Bioproducts Industry: The Technical Feasibility of a Billion-Ton Annual Supply," has been updated twice: once in 2011 ("BT2") and again in 2016 ("BT16"). The updates addressed critical questions such as the cost of biomass, regional differences, contribution of algal fuels, and data reporting.

The Current State of Technology

Ethanol

Food-based ethanol is the most established technology for biofuels in the world (Renewable Fuels Association 2016). It is widely produced and consumed across, and has a prominent place in the U.S. economy. Ethanol has proven to be an effective octane enhancer, giving it a strong market viability. However, expanding ethanol use beyond current levels is controversial and technically challenging.

Cellulosic Ethanol

Cellulosic fuel has lagged behind its peers in the bioenergy technology sphere. Cellulose is found in a wide variety of sources, which presents both opportunities and challenges. Cellulosic ethanol is beginning to be made in industrial quantities.

Algal biofuels

There is great excitement for the emerging technology of using microbial life to produce a variety of petroleum products. Algae have significant potential to support an advanced biofuels industry, especially with maturing synthetic biology technology being employed. However, algal fuels, like cellulosic, have faced severe scale-up challenges that are just beginning to be overcome.

BioProducts

Bioproducts have a great diversity in use and in production method. Many believe bioproducts can serve as a bridge to economic viability for the bioeconomy. This technology has also been helped significantly with the implantation of synthetic biology in the design process.

Codependence of Biofuels and BioChemicals

Biorefining has been identified as a key component of transitioning successfully to a bioeconomy (Lamers, et al. 2016). Biofuels and biochemicals share many of the same logistics such as feedstocks, infrastructure, feedstock commoditization, and land and water use (Bioenergy Technologies Office 2016). They also share many benefits such as carbon sequestration and use, reduced dependence on foreign oil, and growth in local manufacturing.

The benefit of embracing biochemicals are in the economics of production. By volume, approximately 75% of a barrel of crude oil goes towards making fuels, yielding $935 billion in revenue; only 16% of a barrel of oil goes towards making petrochemicals, and despite the smaller volume, these chemicals produce $812 billion in revenue (Bioenergy Technologies Office 2016). Additionally, when looking at market trends, the fuel market is shrinking, while the chemical market is growing at 1% to 1.5%, more than the GDP's annual growth (Bioenergy Technologies Office 2016). Because challenges with both scale-up and economies of scale are hindering biofuels from taking a larger place in the economy, biochemical provide a way to enter the market with lower barriers to entry.

Political and Economic Challenges

The following is an overview of some of the most prominent challenges that are faced and will be faced in creating political and economy sustainability for the bioeconomy.

Current and prospective political climate surrounding RFS

The Renewable Fuel Standard (RFS) has become a polarizing issue for many lawmakers. Its opposition does not fall into neat partisan categories or political ideologies. The key complaints about the Renewable Fuel Standard vary in their merit and consensus, but nonetheless all present challenges to the future of the current legislation. Most notably, these are: the debatable decrease in carbon emissions of current renewable fuels, untimely regulatory action by the EPA, poor estimation of technological progress by the EPA, competition of food-based fuels with the food market, and unintended adverse environmental consequences.

The "Blend Wall"

There has been substantial debate over the amount of ethanol that can and should be blended into the gasoline supply. E10, or a maximum 10% blend of ethanol, is sold extensively throughout the U.S., and accounts for the majority of ethanol sold in the U.S. It is generally agreed that a 10% blend will not harm car engines; however, there is substantial debate as to where this wall actually exits. The volumetric standards set by the EPA will force a higher blend (approximately 10.4%) if the projected fuel consumption for the U.S. is met (US Environmental Protection Agency 2015).

There are 2,795 public fueling stations that sell E85, a blend of gasoline that contains a maximum of 85% ethanol, predominantly in the "corn belt" (US DOE Office of Energy Efficiency and Renewable Energy 2016). This blend of fuel can only be used in Flexible Fuel Vehicles (FFVs). Approximately 17,400,000 FFVs are on the road in the U.S. (US DOE Office of Energy Efficiency and Renewable Energy 2016). In 2016, the EIA estimated that 200 million gallons of E85 were consumed, which is about 0.1% of gasoline consumption (US Energy Information Administration 2016). In fact, most drivers of FFVs do not know that their car is equipped to be fueled with E85, nor do they know where to purchase E85 (US DOE Office of Energy Efficiency and Renewable Energy 2016). Ethanol has a lower energy density than petroleum, and past a certain threshold, its heightened efficiency as an octane enhancer does not balance out (Theiss 2016), therefore E85 is more expensive on an energy content basis. Flex-fuel vehicles running on higher ethanol blends get approximately 25 percent fewer miles to the gallon because of the lower energy content of ethanol (Martin 2016).

Challenges with the LCFS

The legal battle the the LCFS policy faced is far from the only barrier that stood in its way. There are strong opponents to the LCFS who believe that it hurts the economy, and is overly ambitious. Opponents generally believe the LCFS will raise fuel prices, increasing the cost of doing business, and yielding a net negative impact to jobs in California. Various studies by third-parties have concluded that there might be supply concerns to reach the current goal of 10% by 2020. It is being regarded as too aggressive (Fueling California: Consumer Alliance for Responsible Fuel Policies 2012), but the raising cost due to higher carbon fuels will be slightly offset by lower demand of petroleum fuels, as predicted by UC Davis- ITS, (Fueling California: Consumer Alliance for Responsible Fuel Policies 2012) although there is substantial disagreement over how much fuel prices will increase under this policy.

Regulatory Approval Processes: Bioproduct Regulation

The introduction of new products into the market, especially if they are synthesized from genetically modified microbes, can be difficult. Creating the same materials and chemicals that we have produced from unsustainable sources in the new bioeconomy will present regulatory challenges. Corn ethanol faced complications because it was unclear whether it qualified as a part of the food or chemical industry (Bioenergy Technologies Office 2016). New bioproducts face strenuous regulatory challenges with the EPA. It was suggested by various bioproduct researchers and stakeholders that the EPA to work closely with the DOE to reduce barriers to entry (Bioenergy Technologies Office 2016).

Capricious Oil Market

The oil industry has been boom-and-bust for most of its life. As detailed earlier, low oil prices shift economic merits to fossil fuels, while high oil prices spur growth in alternatives such as biofuels. Part of the struggle for biofuels is that the periods of high prices have not been sustained long enough to properly capture the market. It required the confluence of high prices, global instability, and environmental activism to take action in passing a national standard for renewable fuels. This guaranteed the market for alternative fuels, but now that we are again in a period of low oil prices – despite elevated concerns for global insecurity and climate change – biofuels have fallen under harsh scrutiny again. It is likely that this will be a continuing trend, not just because of free market forces and the historic trends of oil markets, but because of the Organization of the Petroleum Exporting Countries (OPEC). Those familiar with the global politics of oil believe that OPEC will potentially flood the market with oil to lower prices and keep out competition with biofuel alternatives.

Young Technology in Bioeconomy

The majority of the strategic initiatives in the National Bioeconomy Blueprint were focused on improving the research and development of biotechnologies, encouraging entrepreneurship, and growing burgeoning industries (Office of Science and Technology Policy 2012). There is no denying that the technology that will be the foundation of this new bioeconomy is still being laid. Federal agencies have been too optimistic in their estimates of emerging fuels, especially those in the "advanced biofuels" category. When using emerging or prospective technologies to solve larger global and political problems such as climate change and energy shortages, politicians are naturally given to overestimating the speed at which a technology can be developed and adopted on a national scale (Breetz 2013).

Strategic Priorities

Economic Stabilization via Bioproducts & Specialization

Some activities that federal agencies (particularly the DOE, USDA, and EPA) can expand upon are:

- Research, development, and demonstration projects for bioproducts
- Collaboration with biorefineries to develop waste utilization
- Devoting resources to hasten bioproduct approval processes

Where possible, federal agencies (particularly the DOE, EIA, and National Laboratories) should:

- Consider markets that will be especially receptive to green technologies when making RD&D investments in bioproduct synthesis
- Expand the presence of and investment in modeling capabilities that predict the market viability for new biotechnology
- Identify and promote technologies that can compete with petroleum alternatives based on performance merits

Target Mutually Beneficial Investments

To grow and support feedstock development, federal agencies (particularly the USDA and DOE) should:

- Further the partnership with farmers to grow localized feedstocks
- Invest in infrastructure that collects, processes, and distributes feedstocks
- Support cogeneration projects in both financing and regulatory measures such as the EPA did for algal fuels in the Clean Power Plan

To promote downstream biorefinery technologies, federal agencies (particularly the DOE) should:

- Further support for biorefinery grant and loan programs
- Partner with the petroleum industry to support conversion of refinery capabilities to co-process biomass

Regional Specialization

One of the most important developments of the Billion Ton Update in 2011 was a regional specialization of analysis of the feedstocks, needs, and abilities that affected the supply curve. This was expanded and refined in the 2016 Billion Ton Update. Regional specialization will be important in identifying what will be most economically advantageous for different companies operating in various areas. Actions that federal agencies (particularly the DOE and USDA) take into consideration are:

- Weigh grant and loan proposals by their merits of leveraging regional advantages and collaborations
- Promote the use and reliance on the Billion Ton Report for business to model the use of biomass in future planning
- Provide networks to collaborate with stakeholders and find regionally specific partnerships

Legislative Proposals

Reform the RFS

End food-based fuel standard. Ethanol has an industry that is robust enough to stand on its own, without a mandate. Ethanol will still be used as an octane enhancer, especially in high performance engines that require high-octane capabilities. Specifically, as there is no corn ethanol mandate, this would constrict the RFS to only overseeing "advanced biofuels," and not allowing for a 20% GHG reduction category. Pushing past the blend wall will need to happen from the consumer side, not from the regulation of the supply side. Mandating more ethanol use should not burden corn producers, but should instead promote advanced fuels. If high ethanol blends such as E85 are to penetrate the market, it should be accomplished with low-carbon cellulosic fuels, if they are to be produced in industrial, economic quantities.

Set proportional – not volumetric – standards. After 2022, Congress should not attempt to set yearly volumetric standards again. New policy should avoid having to predict the transforming transportation industry in setting standards. An updated Renewable Fuel Standard policy should set proportional standards of total fuel consumption for each fuel category. Compliance could still be met with RINs, but there would not be an overall volumetric standard. This would be operationally similar to the LCFS compliance criteria.

Annually update carbon intensity of petroleum fuels. Oil is getting more carbon intensive, and biofuels are getting less carbon intensive. For this reason, there should be an annual update to the carbon intensity metric used to judge the GHG reductions of petroleum. Currently, the EPA uses a 2005 standard. Updating the petroleum standard annually will open up new opportunities to produce fuels that might been on the border of qualifying for as a renewable fuel, or as an advanced biofuel.
This could be opposed by environmental advocates however, who would argue that this allows for biofuels to become dirtier and would not incentivize companies to find cleaner pathways.

Change tax credit for FFVs

The vast majority of people who own and drive a Flexible Fuel Vehicle (FFV) do not know that they are. This has been due to a tax credit being given to automotive manufacturers for each FFV that they sell under the Corporate Average Fuel Economy (CAFE) standards. However, it is my opinion that the tax credit should be on the consumer side, not the supplier side.

I therefore recommend that the CAFE standards should be changed such that the manufacturer no longer receives a tax incentive for producing an FFV. Instead, it should be applied under a program such as the Energy Credit system, maintained by the Department of Energy, that would reward the customer for buying an FFV. Ideally, this would drive demand for FFVs as well as E85 and E15, as more people would be aware of the capability of their car to accept it. This would also serve to accelerate the expansion of stations in the U.S. that sell E85. Ending the tax credit might be opposed by auto industries who benefit from this provision in CAFE standards. It could also face opposition from congressmen and congresswomen in regions that do not have many E85 fueling station and whose constituency would not see the benefits of buying an FFV as readily.

Low Carbon Products Standard

Using the California model of encouraging low carbon fuels, a policy can be formed that will grow the bioeconomy in a market-friendly way. Learning from the challenges faced in the Renewable Fuel Standard, this policy should not try to pick best technologies or to make sweeping market predictions that are subject to much that is out of the control or foresight of the governmental regulators. Instead, a goal should be set to reduce the carbon intensity of chemical products that are entering the market in a Low Carbon Products Standard (LCPS). This can be accomplished either by making a process more efficient, which is always the hope for industry, or by purchasing bioproducts and biochemicals.

The extent to which bioproducts and biochemicals penetrate the market is correlated to the expansion of market viability for the bioeconomy. I propose that a conservative inaugural list of applicable chemicals would be set forth in the first years of implementation that have a demonstrable ability to be supplemented by biological alternatives (e.g. benzene, aldehydes, fatty acids). The applicable chemicals should be broad in their usage – "platform" chemicals, that are further refined in product manufacturing, should be regulated, rather than specific end-products. Compliance would be overseen by the EPA, and the qualifying LCPS chemicals would be expanded and contracted as needed by the EPA. Obligated parties would be companies who purchase chemicals, and companies could "opt-in" if they generate renewable chemicals and can sell their products and credits in the market. The precipitating effects of this policy would be beneficial for the growth of the bioeconomy overall. Although biochemicals are produced in vastly smaller quantities compared to biofuels, a growth in one industry can implicitly help the other. This would help grow the establishment of feedstocks, expand research and development in critical biological engineering areas, and help build economic solvency for bio-based industries.

Price Carbon

A carbon fee and dividend (CFD), as proposed by the bipartisan group, Citizens Climate Lobby (CCL), would work by pricing carbon at the source of entry to the market, such as at the well, pump, mine, or border (Citizens Climate Lobby 2016). The pricing would start low and would increase steadily over time, establishing predictability in the market. All revenue generated from the carbon fee would be returned on an equal basis back to families, according to the size of the household (except for minor administrative expenses) (Citizens Climate Lobby 2016). This fee model does not grow the government, and still allows for signors of the Norquist pledge to abide by their promise of not raising taxes.

According to an economic analysis by the third party institution, Regional Economic Models, Inc. (REMI), the policy as modeled will add 2.8 million jobs to the economy over 20 years, and will meet its goals in carbon reductions (Citizens Climate Lobby 2014). A policy such as this will be instrumental in engendering investments in biotechnology, because of its natural carbon sequestering capabilities. It can also foster political sustainability by building an important bipartisan collaboration to address a critical issue in a way that appeals to the values of both parties. A policy such as this would disrupt the economy and would hurt jobs in industries that are reliant on fossil fuel production. The REMI industry that while there is a net gain in jobs to the economy nationally, some regions experience a net loss (Citizens Climate Lobby 2014). An economic disruption such as this is very difficult to predict (Daniels 2016), and unforeseen complications are expected.

Summer 2016

Power Up: Advancing the Automotive Industry through Energy Storage

Mena-George R. Basaly

Washington Internship for Students in Engineering, WISE

Sponsored by the American Institute of Chemical Engineers, AIChE

WISE Washington Internships for Students of Engineering

AIChE The Global Home of Chemical Engineers

EXECUTIVE SUMMARY

This report proposes policy recommendations to promote innovative storage technologies with application to Electric Vehicles (EVs) and Plug-In-Hybrid-Vehicles (PHEVs). Federal, local, and state governments should take the following measures to integrate and normalize the EV/PHEV market:

1. Reform the $7,500 tax incentive in place for current EV/PHEV owners and modify it to a point-of-sale rebate for consumers.

2. Implement a 5-year-federal-tax-free period for startups that are developing the EV battery market to facilitate innovation of inexpensive, long-range batteries.

3. Encourage and protect consumer usage of EVs through raised standards and incentives.

While the renewable energy storage market has grown from $111 million in 2013 to $441 million in 2015, it is expected to exceed $2.9 billion by 2021. This growth in innovation is not proportional to fuel use in the transportation sector. In 2014, fossil fuels made up 92% of the total energy consumed by the U.S. transportation sector, in comparison to 60% of total U.S. energy consumption across all sectors. Thus, these policies will spur growth to achieve EV market integration and surpass President Obama's goal of 1 million EVs on the road by 2015[1].

A point-of-sale rebate of up to $10,000 should replace the $7,500 tax incentive. This incentive was proposed in 2009 as part of the American Recovery and Reinvestment Act during the Great Recession in order to stimulate EV growth. However, the policy is outdated and serves as a tax refund for small batteries (under 20 kWh). This new policy will benefit middle class consumers

---

[1] The U.S. reached about 400,000 EVs on the road by the end of 2015.

who do not have the money upfront, as well as encourage manufacturers to drive down EV costs and begin fading out the point-of-sale refund.

A 5-year-tax-break for startups would allow innovation on the university level and bridge the gap between academia and industry. A similar pilot program was launched in New York, *START UP NY*, and it proposed a 10-year-tax-free period for startups working at or near partnering universities. This program costs the state of New York $323 million. Thus, if implemented on a federal level, this could reach $1 billion. Reducing federal oil subsidies, which amounted to $37.5 billion in 2014, can easily offset this cost.

Encouraging and assuring consumers is necessary to make EVs prevalent and competitive with internal combustion vehicles. Regulatory standards, financial incentives, and non-financial incentives governing tail pipe emissions, parking privileges, car pool lane access, and toll reductions are few of many policies that could encourage consumer acceptance. Much of the drawback of integrating EVs to the market is the lack of consumer awareness, and consumer uncertainty about safety, range, and cost. In order to combat such concerns, these policy alternatives could be introduced, in addition to state and local incentives.

The bottom line is that electric vehicles are here to stay. These groundbreaking, clean technologies, along with fuel cells and other zero-emission technologies could potentially be the future of the automotive industry. Policy reform of tax incentives, startup incentives, and consumer protection and encouragement will foster innovation, cost reduction, energy security and independence in the United States, and long-rang targeting that pose a viable alternative to traditional internal combustion engine.

TABLE OF CONTENTS

PREFACE

ABOUT THE PROGRAM

The Washington Internship for Students in Engineering (WISE) was founded in 1980 through the collaborative efforts of several professional engineering societies to encourage and prepare future leaders in the engineering realm to contribute to increasingly important issues at the intersection of science, technology, and public policy.

Each summer, the WISE societies select outstanding engineering students from a nation-wide pool of applicants. The students spend nine (9) weeks living in Washington, D.C. during which they learn how government officials make decisions on complex technological issues, and how engineers can contribute to the legislative process and regulatory public policy decision-making.

ABOUT THE AUTHOR

Mena-George Basaly graduated from the City College of New York in June 2016 with a Bachelor's degree in Chemical Engineering and concentrations in economics and mathematics. At CCNY, he was heavily involved in the AIChE Student Chapter. He has previously held internships at SkyBridge Capital and MBS Value Partners to gain exposure to the world of finance and Wall Street, and then ventured into the tech world to intern at Global Futures Group, a smart cities startup that helped deepen his interest in clean energy. After the WISE Program, he will launch his career at Cognizant Technology Solutions as a Technology Consulting Analyst.

ACKNOWLEDGEMENTS

I would like to thank the American Institute of Chemical Engineers (AIChE) for sponsoring me this summer, particularly Steve Smith and Dr. Marshall Lih for their guidance and support. I would also like to thank the faculty member-in-residence, Dr. Michael Marcus, for scheduling an amazing series of speakers this summer. I owe a special thanks to IEEE for loaning me a cubicle, and Erica Wissolik for being available and providing support. Another set of thanks go to the Department of Chemical Engineering at City College, particularly Dr. Robert Messinger for his guidance and support on battery technology background and help in shaping this paper. I would also like to thank General David Petraeus and Dr. Elizabeth Biddinger for their generous recommendations. Finally, I would like to thank my mentor, Dr. Dale Keairns, for his help and invaluable feedback.

LIST OF TABLES AND FIGURES

LIST OF ACRONYMS

AC	Alternating current
APPA	American Public Power Association
BTU	British thermal unit
DC	Direct current
DOE	Department of Energy
EERE	Energy Efficiency and Renewable Energy
EIA	Energy Information Administration
EPA	Environmental Protection Agency
EV	Electric vehicle
GHG	Green house gas
IRENA	International Renewable Energy Agency
kWh	Kilowatt-hour
Li-ion	Lithium Ion
Na-S	Sodium Sulfur
Ni-Cd	Nickel Cadmium
PHEV	Plug-in-hybrid-electric-vehicle
PV	Photovoltaic
RPS	Renewable Portfolio Standards
SBIR	Small Business Innovation Research
STTR	Small Business Technology Transfer

INTRODUCTION

The renewable energy storage market has grown from $111 million in 2013 to $441 million in 2015 and is expected to exceed $2.9 billion by 2021 **[1]**. Renewable energy storage has propelled incredible advances in battery technology, which has allowed the industry to grow at such a rapid pace. The International Renewable Energy Agency (IRENA) expects the global share of renewables to double by 2030 **[2]**. This expansion means a rapid increase in electric vehicles (EVs) and plug-in-hybrid vehicles (PHEVs).

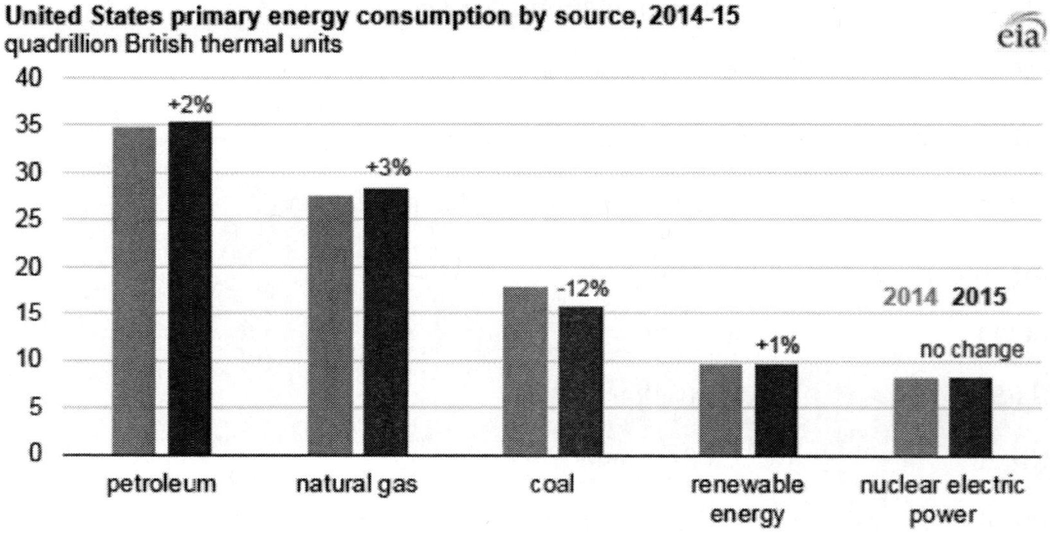

Figure 1: US energy consumption by source from 2014 to 2015

According to the United States Energy Information Administration (EIA), the United States consumes approximately 10 quadrillion British thermal units (BTUs) of renewable energy, which is significantly less than petroleum and natural gas, as portrayed in **Figure 1.** Coal provided 16% of the total energy use in the US, down from 18% in 2014.

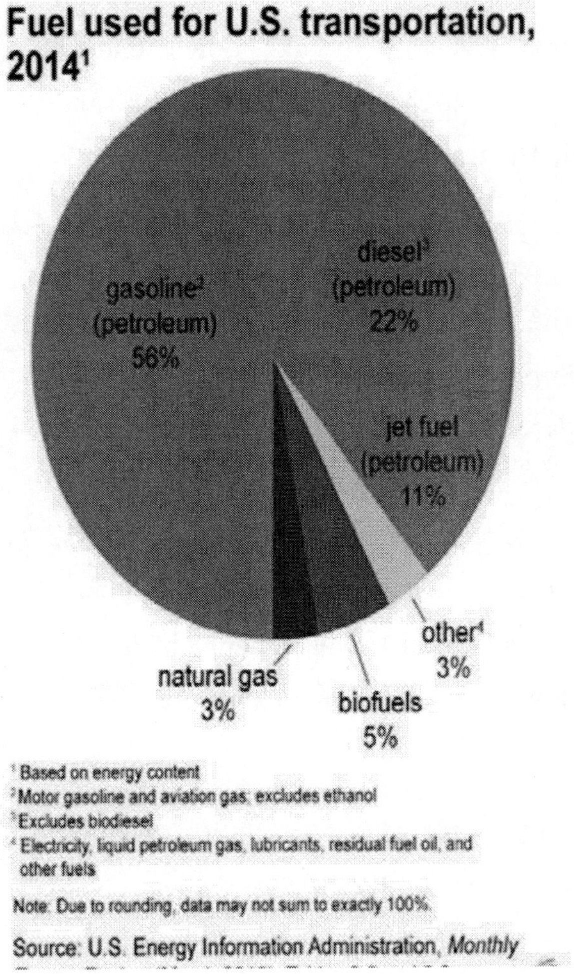

Figure 2: Energy consumption by fuel type in the transportation sector in 2014

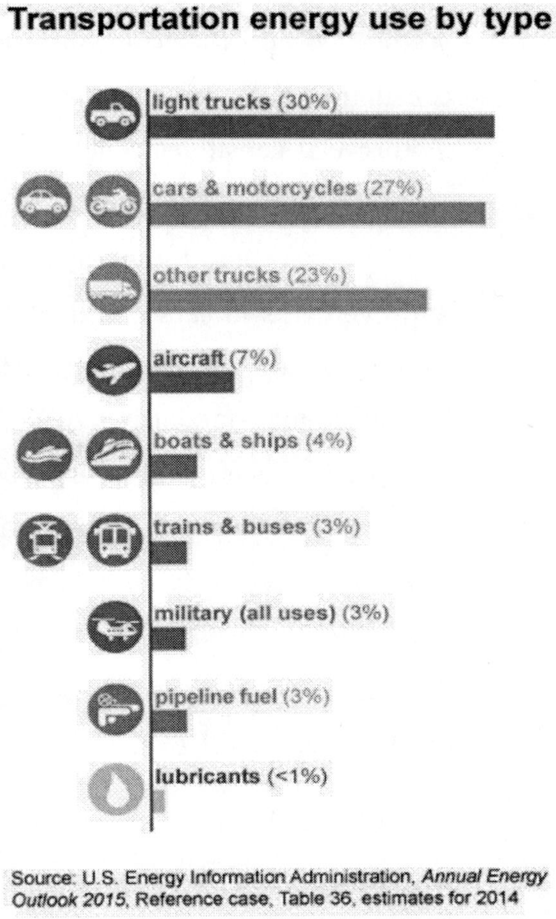

Figure 3: Energy consumption by mode of transportation in 2014

Natural gas increased to 29% in 2015, an increase larger than that for any other fuel. In the transportation sector, electrical batteries are almost miniscule.

Figure 2 indicates that less than 3% of the transportation sector was powered by electrical means, while 56% relied on gasoline in 2014.

Petroleum fuels are made from crude oil and liquids from hydraulic fracturing of natural gases. The petroleum fuels used for transportation include gasoline, diesel fuel, jet fuel, residual fuel oil, and liquid petroleum gases. In 2014, those fuels made up 92% of the total energy consumed by the U.S. transportation sector. Within the transportation sector, Figure 3 breaks down energy consumption by mode of transportation. In 2014, light trucks[2], car, and motorcycles consumed roughly 60% within the transportation sector.

U.S. transportation energy use has increased, despite fuel economy improvements in cars and light trucks. The national average fuel economy for passenger cars and for light trucks has improved over time mainly because of fuel economy standards set by the federal government for those types of vehicles. However, total transportation fuel consumption has generally increased because of an increase in the number of vehicles (especially light pickup trucks, sport utility vehicles, and heavy duty freight trucks). Fuel consumption has generally increased because of the number of miles traveled per vehicle **[2]**. Thus, given the increase in fuel consumption, now is the time to shift the market towards EVs and PHEVs.

This report discusses the importance of EVs and PHEVs. In 2008, President Obama set a goal to have one million EVs on the road by 2015. However, given high battery costs required for the necessary electrical energy storage, the market penetration of EVs has been quite low. In fact, by the end of 2015, there were roughly 400,000 EVs on the road, with 25% of them in California alone **[3]**.

---

[2] A light truck is defined as a truck with a payload capacity of less than 4000 lbs (according to Title 40 of the Code of Federal Regulations created by the EPA).

To be competitive, policies and incentives must be put in effect for both EVs and PHEVs, particularly to bring down the cost of batteries.

BACKGROUND

Given the current Energy Revolution in North America and the advancement of renewable energy, the array of storage technologies is rapidly growing to accommodate these renewables.

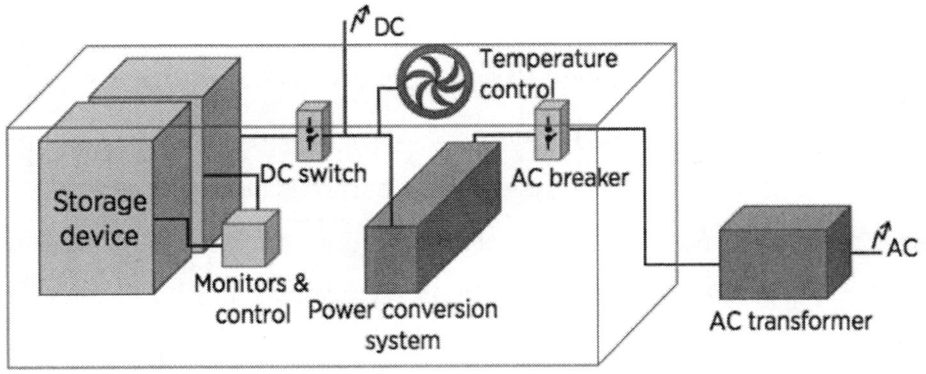

Source: Based on EPRI and DOE, 2013

Figure 4: Battery Storage System with its primary power components

A battery storage system contains several primary components as shown in **Figure 4**, including the battery, monitoring and control systems, and a power conversion system. Cell based batteries consist of individual cells connected into modules and then into packs. Flow batteries consist of external tanks filled with an electrolyte, which flows through a reaction stack. Monitoring and control systems, referred to as the battery management system, ensure safety and maximize performance. The battery management system prevents individual cells from overcharging, and controls charge and discharge of the battery. This is important for safety and performance. Battery cells and component monitoring may vary to some degree, in that different types require emphasis on particular issues. For instance, lithium-ion battery packs must emphasize thermal monitoring and controls, given a tendency to overheat[3].

TYPES OF BATTERIES

A battery, by definition, is a device consisting of one or more electrochemical cells that convert chemical energy into electrical energy. The positive terminal of the cell is the cathode and the negative terminal is the anode. Electrolytes allow ions to move between the terminals, and current then flows through the battery to do work, and provide electrical energy.

Lithium Ion (Li-ion) batteries [10] are very prevalent in today's society. They are used in electric vehicles, cell phones, laptops, photovoltaic (PV) applications, and even megawatt (MW) containerized batteries for grid storage. Li-ion batteries are categorized by the transfer of lithium ions between the electrodes during the charge and discharge reactions. Typically, li-ion batteries have high cell voltages (3.6 V to 3.7 V), which correspond to high energy density.

Lead-acid batteries [11] are high power, inexpensive, safe, and reliable. However, low specific energy, poor cold-temperature performance, and short calendar and cycle life are the main barriers to entry. Advanced high-power lead-acid batteries are being developed, but these batteries are only used in commercially available EVs for ancillary loads.

Nickel Metal Hydride (Ni-MH) batteries [11] offer realistic specific energy and specific power capabilities. Ni-MH batteries have a much longer lifecycle than lead-acid batteries and are safe and cannot be abused, which makes them great for EVs. However, the main challenges with Ni-MH batteries are their high cost, high self-discharge and heat generation at high temperatures, and the need to control hydrogen loss.

---

[3] Hence the recent backlash with hover boards.

Ultracapacitors [11] are storage devices that store energy in a polarized liquid between an electrode and an electrolyte. To be clear, ultracapacitors are not batteries, but are highly used in the EV/hybrid vehicle industry. They increase storage capacity as the liquid's surface area increases. Ultracapacitors are useful in offering more power to vehicles during acceleration and hill climbing, as well as recycling and recovering braking energy. They may also be useful as secondary energy-storage devices in electric drive vehicles because they help electrochemical batteries level load power.

TYPES OF ENGINES

INTERNAL COMBUSTION ENGINE (ICE)

Combustion is a basic chemical process that releases energy by essentially burning fuel. The ignition and combustion of the fuel occurs inside the ICE, which then converts energy to work to power the engine. An ICE is a piston/cylinder, which is simply a reciprocating engine. The burned fuel creates gases that push the piston to rotate the crankshaft. The crankshaft then powers a series of gears to then drive and power the vehicle **[16]**.

The two prevalent kinds of ICEs are spark ignition gasoline engines and compression ignition diesel engines (simply gasoline vs. diesel vehicles). Most of these are four-stroke cycle engines, meaning four piston strokes are needed to complete a cycle **[16]**. The four components of the engine process include intake, compression, combustion and power stroke, and exhaust.

Gasoline and Diesel engines differ in how they handle fuel. In a gasoline engine, fuel and air are mixed and inducted into the cylinder during the intake process. After the piston compresses the mixture, the spark combusts the mixture of gases. The expansion of the combusted gases powers the cylinder during the power stroke phase. In a diesel engine, only air is inducted into the engine and then compressed. Diesel engines spray hot compressed airs with the fuel to ignite and combust it **[16]**.

ELECTRIC MOTOR

EVs use electricity stored in a battery pack to power an electric motor and turn the wheels. When depleted, the batteries are recharged using grid electricity, either from a wall socket or a dedicated charging unit. Since they don't run on gasoline or diesel and are powered entirely by electricity, EVs do not have any tailpipe emissions **[19]**.

The battery is the main component of an electric vehicle. The controller is the connecting factor between battery and the motor, monitoring how much energy is transmitted. When the driver presses down on the gas pedal, the controller sends the power to the engine to accelerate. The controller in EVs is either operating on alternating current (AC) or direct current (DC) **[20]**.

With an AC controller, the DC voltage from the battery is pulsed through the system. A typical AC controlled vehicle motor operates on 240 volts. A DC controlled vehicle operates from 96-192 volts **[20]**.

CURRENT LEGISLATION

CLEAN ENERGY INVESTMENT INITIATIVE

The White House passed a Clean Energy Investment Initiative **[3]** to reduce carbon emissions. President Obama announced a $4 billion goal to catalyze private sector investment in solutions to climate change, including innovative technologies with breakthrough potential to reduce carbon pollution. These technologies include solar voltaics, advanced batteries, lighting, and fuel cells.

EXECUTIVE ACTIONS TO INCREASE CHARGING INFRASTRUCTURE

In July 2016, the Obama Administration announced an unprecedented coalition to add and deploy more EV charging infrastructure, which will add more EVs on the road. This coalition includes the Federal government, private sector, and states, as well as a new framework for collaboration for vehicle manufacturers, electric utilities, electric vehicle charging companies, and states. President Obama's plan highlights a $4.5 billion loan guarantee from the DOE in order to invest in charging station infrastructure. The FAST Act is also part of the plan, which will identify zero emission opportunities for electric vehicle charging across the country. The Administration has an ambitious, yet accessible, goal to develop a national network of EV infrastructure by 2020.

Additionally, the Office of Energy Efficiency and Renewable Energy (EERE), a subset of the DOE, will bring forth a Memorandum of Understanding (MOU) with the American Public Power Association (APPA) to collaborate on EV infrastructure and access in municipalities across the United States. EERE and APPA will provide information to increase education and awareness of the benefit of EVs to public power utilities and local officials, and develop a community action plan focused on smaller communities with fewer than 200,000 electric customers. The partnership will also work to enhance workplace charging efforts at public power utilities, study the impacts of EVs in public power communities, and share insights regarding infrastructure installation and EV interaction with the modern grid.

CLEAN ENERGY INVESTMENT CENTER

The Department of Energy (DOE) launched a Clean Energy Investment Center to focus on four goals:

1. **Single Point of Access for Information** to allow investors to access clean energy reports and companies funded by the DOE.

2. **Technical Assistance for National Laboratories** to share research and analysis produced by DOE and its 17 national laboratories on relevant developments in clean energy technology.
3. **Information on Early-Stage Projects and Companies** including the Advanced Research Projects Agency-Energy (ARPA-E), Small Business Innovation Research (SBIR), Small Business Technology Transfer (STTR), and others that help fund and accelerate emerging early-stage technology projects and companies. The center will aggregate and make available public information on entities currently engaged in partnerships with DOE.
4. **Connections to Additional U.S. Federal Government Programs** that include information about energy and climate programs at other government agencies including the U.S. Department of Agriculture, U.S. Department of Housing and Urban Development, U.S. Department of Transportation, U.S. Department of Treasury, U.S. Environmental Protection Agency, U.S. Small Business Administration and the National Science Foundation.

CLEAN ENERGY TAX INCENTIVE

Additionally, Senator Martin Heinrich (D-NM) **[4]** will introduce a bill that provides a clean energy tax incentive. The legislation would give businesses and homes a 30% credit, but the credit would taper off starting in 2020. The bill would create incentives for both business and residential deployments. The business incentive applies to systems of at least 5 kilowatt-hours from any of the available technologies, including batteries, flywheels, pumped hydro, thermal energy and compressed air. The credit only applies to battery storage for residential applications, and the systems must have at least 3 kilowatt-hours of capacity. For comparison, the Tesla Powerwall has 6.4 kilowatt-hours of capacity.

H.R. 8 NORTH AMERICAN ENERGY SECURITY ACT AND INFRASTRUCTURE ACT OF 2015

A section of H.R.8 aims to amend the Energy Policy Act of 2005 to require electric utilities to make an interconnectivity service and net metering available to an electricity consumer for community solar facilities, consisting of a solar PV system. These facilities have a panel rating of 2 megawatts or less and allocate electricity to multiple consumers as well as meeting other standards. H.R.8. also states that each state regulatory commission is allowed to set rate making standards for non-regulated utilities.

S.2012 NORTH AMERICAN ENERGY SECURITY AND INFRASTRUCTURE ACT OF 2016

Section 3114 of S.2012 is a battery storage report that poses the following questions regarding the outlook for the storage industry:

1) How do existing Federal standards impact the development and deployment of battery storage systems?
2) What are the benefits of using existing battery storage technology, and what challenges exist to their widespread use? What are some examples of existing battery storage projects providing these benefits?
3) What potential impact could large-scale battery storage and behind-the-meter battery storage have on renewable energy utilization?
4) What is the potential of battery technology for grid-scale use nationwide?
5) What is the potential impact of battery technology on the national grid capabilities?
6) How much economic activity associated with large-scale and behind-the-meter battery storage technology is located in the United States? How many jobs do these industries account for?

7) What policies other than the Renewable Energy Investment Tax Credit have research and available data shown to promote renewable energy use and storage technology deployment by State and local governments or private end-users?

These questions serve as a backbone to what's left to address in the storage industry.

H.R.4584 ELECTRIC CARS ACT

Electric Credit Access Ready at Sale Act of 2014 **[8]**, or the Electric CARS Act, amends the Internal Revenue Code to increase the maximum amount of the tax credit for new qualified PHEVs to $7,500 and to extend the termination date of such credit to December 31, 2020. It permits the assignment of the credit to the sellers of such vehicles. However, this bill has only been introduced, and has not seen any activity since its introduction in May 2014.

DISADVANTAGES

Many of the hurdles to common usage of EVs/PHEVs over the past years include range on a full charge and car costs, both of which are likened to the high cost of advanced battery technologies. However, **Figure 5** illustrates the associated projected and current costs and energy densities of EV batteries. Costs have consistently gone down year-to-year, while energy densities have gone up year-to-year. While costs are still high, with the right policies, these costs can significantly decrease further. Additional barriers relate to the need to access recharging facilities, costs and other challenges associated with their installation, as well as the lack of awareness or confidence in the technology **[9]**. However, the Obama administration recently allocated over $4 billion to the DOE in order to invest in infrastructure for charging stations.

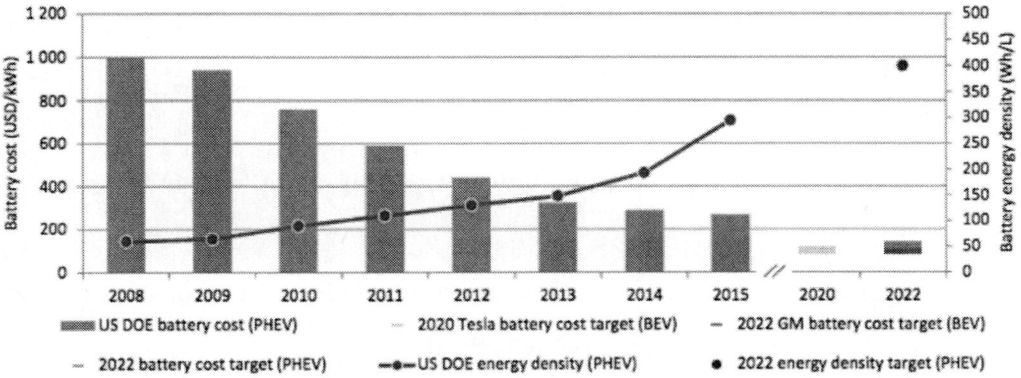

Notes: USD/kWh = United States dollars per kilowatt-hour; Wh/L = watt-hours per litre. PHEV battery cost and energy density data shown here are based on an observed industry-wide trend, include useful energy only, refer to battery packs and suppose an annual battery production of 100 000 units for each manufacturer.

Sources: US DOE (2015 and 2016) for PHEV battery cost and energy density estimates; EV Obsession (2015); and HybridCARS (2015).

Figure 5: Target and actual costs and energy densities of EV/PHEV batteries.

POLICY RECOMMENDATIONS

1. **Reform the $7,500 tax incentive for EVs/PHEVs to accommodate the middle class [6].**
   This report proposes a reform of the tax incentive laid out in the Internal Revenue Code (IRC) 30D. Instead of consumers waiting a year (or more) to receive the rebate and pay money up front, it is imperative that this incentive is actually a point-of-sale refund for consumers up to $10,000.

   The current policy in place for EVs and PHEVs is a tax rebate ranging from $2,500 to $7,500 **[5]**. According to the American Recovery and Reinvestment Act, "the base tax credit is $2,500 and above 5 kilowatt-hours taxpayers may claim on their IRS return $417 per kilowatt-hour not to exceed $5,000." This is where the "$7,500" tax credit comes from. However, cars with bigger batteries like the 24-kwh Nissan Leaf or 60-85-kwh Tesla Model S still are only eligible for the $7,500 tax credit, which isn't attractive to consumers **[12]**. Further, each manufacturer is capped at a selling limit of 200,000 PHEVs it could sell in the U.S., and this tax credit applies only to leases.

In order to implement this policy, the following measures must be addressed:

- Raising the federal fuel excise tax to offset the cost of the newly implemented point-of-sale refund.
- Manufacturers would administer the refund, and then apply for reimbursement from the federal government. This will, in the long run, drive down the upfront cost of the EVs and PHEVs.

2. **R&D Tax incentive for early stage energy storage developers**

This report proposes piloting a tax-free operation for battery startups that focus on EV/PHEV battery applications. This program would be a five-year federal tax break for startup battery developers and researchers in order to stimulate less expensive, long-range batteries that can power EVs and PHEVs. Moreover, the five-year-tax-break can be linked to a mandatory kWh goal, set by the EPA or DOE, that these startups have to meet.

This incentive is modeled after Start Up NY, which states: *"START-UP NY offers new and expanding businesses the opportunity to operate tax-free for 10 years on or near eligible university or college campuses in New York State."* **[7]**

While START-UP NY costs New York State $323 million **[15]**, this federal incentive would reach the billions. Thus, while this sounds like an expensive policy, this report proposes an off-setting decrease in oil subsidies. According to *Oil Change International*, as of July 2014, U.S. fossil fuel subsidies reached $37.5 billion annually, including $21 billion in production and exploration subsidies **[14]**. While these subsidies are justified by increasing energy security in the United States, the oil industry is mature and has made significant advances in exploration, particularly in hydraulic fracturing.

The benefits of implementing a five-year-tax-free plan would include the following:

- Increased United States Energy Security by diversifying energy storage, given the volatile nature of oil and gas markets and global political instability
- Spurring innovation at the university level and bridging the gap between industry and academia. This will give researchers (at Stanford's Energy Labs, City University of New York's Energy Insistute, MIT's start-up hub, etc.) the opportunity to apply their research to industry and accelerate innovation in battery storage to drive down costs and increase ranges to be comparable to internal combustion engines.

3. **Raise standards to protect and encourage consumer usage of EVs/PHEVs**

Much of drawback of integrating EVs into the market is the lack of consumer awareness, and consumer uncertainty about safety, range, and cost. According to report published by Accenture **[21]**, vehicle range control, and charging infrastructure are main consumer concerns. In order to combat this, a policy alternative could be to introduce climate-friendly standards, in addition to state and local incentives to promote consumer purchase and usage of EVs/PHEVs.

For example, the Netherlands allows road tax exemption for zero-emission cars. In 2015, this also applied to PHEVs emitting less than 50 g CO2/km. In 2016, they are subject to charges that are only 50% of the road tax for an internal combustion engine car **[13]**. Other countries in

Europe have proposed similar incentives for consumers. This report suggests the following policy alternatives with these benefits:

- Regulatory standards that regulate tailpipe emissions and fuel economy standards
- Financial incentives that include the following[4]:
  - Consumer access to restricted lanes (traditionally carpool lanes)
  - Parking preferences in municipal lots, garages, and meter-free parking in metered lots
  - Toll reductions: the best way to introduce this is a system like the Netherlands. Highway officials can offer up to 75% discounts on tolls for EVs/PHEVs, which will reduce emissions in congested areas with expensive tolls (i.e. Tri-State area)

CONCLUSION

Given the current state and legislation of battery technology and EVs, electric vehicles can potentially be commonplace in the global automotive market. Whether or not they do lead the auto industry depends on the future of policy surrounding EVs and battery technology—particularly government subsidies and incentives.

Policy reform of tax incentives, combined with policy proposals for startup incentives, and consumer protection and encouragement will foster innovation cost reduction, energy security and independence in the United States, and long-range targeting that will pose a viable threat to the traditional internal combustion engine.

[4] These financial incentives are on the State and Local levels, in collaboration with the Department of Transportation.

REFERENCES

[1] (GTM Research, 2016)

[2] (U.S. Energy Information Administration, 2015)

[3] (Shepardson, 2016)

[4] (IRENA, 2015)

[5] (DOE, n.d.)

[6] (Spector, 2016)

[7] (EV Wiki, n.d.)

[8] (Cob, 2015)

[9] (StartUp NY, n.d.)

[10] H.R.4584

[11] (International Energy Agency, 2016)

[12] (Energy Storage Association, 2016)

[13] (EERE, 2016)

[14] (Cob, 2015)

[15] (International Energy Agency, 2016)

[16] (Oil Change International, 2016)

[17] (Coin, 2013)

[18] (EERE, 2013)

[19] (Union of Concerned Scientists, 2014)

[20] (CarsDirect, 2012)

[21] (Accenture, 2014)

BIBLIOGRAPHY

Accenture. (2014). *The Electric Vehicle Challenge.* New York: Accenture. Retrieved from Accenture Publications: https://www.accenture.com/mz-en/~/media/Accenture/Conversion-Assets/DotCom/Documents/Global/PDF/Industries_15/Accenture-Electric-Vehicle-Challenge.pdf

CarsDirect. (2012, Mar 14). *How an Electric Car Motor Works.* Retrieved from CarsDirect: http://www.carsdirect.com/green-cars/how-an-electric-car-motor-works

Cob, J. (2015, March 6). *Does the $7,500 Plug-In Car Tax Credit Need Reform?*. Retrieved from HybridCars: http://www.hybridcars.com/does-the-7500-plug-in-car-tax-credit-need-reform/

Coin, G. (2013, Aug 6). *NY Minute: Tax-free Start-Up NY program will cost $323 million, Gov. Andrew Cuomo's budget office says.* Retrieved from Syracuse: http://www.syracuse.com/news/index.ssf/2013/08/gov_andrew_cuomo_tax-free_start-up_ny_program_will_cost_323_million.html

DOE, U. (n.d.). *U.S. Department of Energy's Clean Energy Investment Center.* Retrieved from http://energy.gov/technologytransitions/us-department-energys-clean-energy-investment-center

EERE. (2013, Nov 22). *Internal Combustion Basics.* Retrieved from EERE: http://energy.gov/eere/energybasics/articles/internal-combustion-engine-basics

EERE. (2016, May 10). *Batteries for Hybrid and Plug-In Electric Vehicles.* Retrieved from Alternative Fuels Data Center: http://www.afdc.energy.gov/vehicles/electric_batteries.html

Energy Storage Association. (2016). *Lithium Ion (LI-ION) Batteries.* Retrieved from ESA: http://energystorage.org/energy-storage/technologies/lithium-ion-li-ion-batteries

EV Wiki. (n.d.). *Tax Incentives, Rebates*. Retrieved from Electric Vehicle Wiki: http://www.electricvehiclewiki.com/Tax_Incentives,_Rebates

GTM Research. (2016, June 01). *U.S. Energy Storage Monitor: Q2 2016.* Retrieved June 27, 2016, from Green Tech Media:

https://www.greentechmedia.com/research/subscription/u.s.-energy-storage-monitor

International Energy Agency. (2016). *Global EV Outlook 2016*. Retrieved from IEA: http://www.iea.org/publications/freepublications/publication/Global_EV_Outlook_2016.pdf

IRENA. (2015, January). *Battery Storage for Renewables: Market Status and Technology Outlook*. Retrieved from International Renewable Energy Agencey.

Oil Change International. (2016). *Fossil Fuel Subsidies: Overview*. Retrieved from Price of Oil: http://priceofoil.org/fossil-fuel-subsidies/

Shepardson, D. (2016, Jan 20). *Electric vehicle sales fall far short of Obama goal* . Retrieved Aug 03, 2016, from Reuters: http://www.reuters.com/article/us-autos-electric-obama-insight-idUSKCN0UY0F0

Spector, J. (2016, July 5). *Senator Heinrich to Introduce Energy Storage Tax Credit Bill Next Week* . Retrieved from GreenTech Media.

StartUp NY. (n.d.). *Business Growth*. Retrieved from StartUP NY: https://startup.ny.gov/business-growth

U.S. Energy Information Administration. (2015, Sept 23). *Use of Energy in US Explained: Energy Use for Transportation*. Retrieved Aug 3, 2016, from US EIA: http://www.eia.gov/Energyexplained/?page=us_energy_transportation#tab2

Union of Concerned Scientists. (2014). *How Do Battery Electric Cars Work?* Retrieved from UCSUSA: http://www.ucsusa.org/clean-vehicles/electric-vehicles/how-do-battery-electric-cars-work#.V88IH5iV_Dd

475209 Challenges to the Partnership between Research and Education

Abstract:

The current model for a "professor" of engineering at a Carnegie research university assumes the professor is a scholar, teacher, advisor and university citizen. The financial underpinning of this model rests on revenue generation through tuition (teaching credits), research grants and endowment draw to various degrees depending on the institution and its resources. Several factors are challenging this model: resistance to increases in tuition rates, reduced success rates on proposals and more university cost sharing, more demands on teaching at the undergraduate level, federally mandated compliance, and tenure itself. Unless research is more tightly connected to education, it will become more difficult to sustain the "professor" model in its classical sense.

475208 The Next Gen Engineers

Abstract:

This is a pivotal moment in time for the engineering workforce. There are demographic gaps, pressures for more rapid innovation and knowledge transfer and acquisition, complex multi-national companies and global economies, desire for a more diverse workforce, and the need to increase the awareness of the impact of engineers on society. We need to excite, attract, retain, and prepare the next generation of engineers to be innovators, entrepreneurs, professional ambassadors, leaders, and to be resilient and opportunistic. We also need to recognize that the potential pool of next-generation engineers views the world through lenses shaped by their personal experiences and learning styles, and that their motivations, professional interests, and career trajectories might be very different from the past. These brief introductory comments will summarize recurring themes and reactions to discussions with industry leaders, late-, mid-, and early-career engineers, legislators, faculty, and students entering the engineering pipeline.

475210 The Changing Face of Public Higher Education

Abstract:

Public higher education is at a crossroads. The dramatic decisions made by states during and soon after the Great Recession drove extraordinary increases in tuition and, in many cases, a spike in student debt. The question before us is how to continue to deliver on our missions of teaching, research and public service in the presence of ongoing budget challenges. Options to be explored included public-private partnerships, educational efficiencies, and a refocusing of priorities, with an emphasis on the STEM fields.

475517 Engineering Nanocrystalline Photocatalysts for Energy and Environmental Applications

Abstract:

Nanoparticles and nanocrystals of semiconducting oxides constructed from nanoscale building blocks often possess unique and much improved properties. Such materials are promising in enabling innovative technologies for conversion and storage of renewable energies for the future. With current challenges in climate change and sustainable development, nanotechnology is especially exciting because it provides great opportunities for technological advances in areas of solar power, solar hydrogen production and storage.

This lecture presents latest developments in oxide nanomaterials such as titania as photocatalysts. Materials such as layered titania and single anatase crystals are photoactive materials promising for cheap and efficient solar cells, hydrogen production from water splitting and solar detoxification of water and air. Latest advances in bandgap engineering of TiO_2 for visible light photocatalysis will be highlighted and discussed.

472829 Poster Presentation Success: How to Prepare and Present a Winning Poster

Abstract:

Preparing a poster presentation provides a great opportunity to connect with potential collaborators and employers. The key to success is optimizing your time before, during and after the session, and really taking advantage of the eyeballs on your poster to build a strong network and develop job leads. In this workshop, weâ€™ll discuss the elements of a successful poster and how to deliver a meaningful presentation. You will learn how to craft your value statement and talking points, and how to effectively engage those who stop by. Youâ€™ll also discover how to utilize social media to build buzz around you and your research endeavors, and how to follow up on the connections you make. Speaker: Alaina G. Levine, President, Quantum Success Solutions, Author: Networking for Nerds (2015 Wiley)

466412 Electrophoretic Separation of Biomolecules in Nano-Lipid Membrane

Abstract:

The high performance separation of membrane proteins has not yet been attained in despite of the significance, and been demanded. For that, we developed a novel separation method in a nano-lipid membrane, which utilizes the generated function of nano-lipid membranes.The nano-lipid membranes were constructed on the glass slide by self-assembly of lipid molecules. At first we investigated the factors affecting the separation in the nano-lipid membrane. In the nano-lipid membrane, membrane proteins kept its original structures as like in the cell membranes. In addition, the diffusion of membrane proteins was suppressed compared that in free solution. These specific properties of the nano-lipid membrane enable non-denaturing separation and high separation efficiency. Furthermore, we clarified that the separation efficiency of membrane proteins could be easily controlled by varying the composition and phase state of lipid membrane. Finally, we attempted the electrophoretic separation of membrane proteins and easily attained it in the nano-lipid membrane. Thus, we could successfully demonstrate the possibility of high performance separation in nano-lipid membranes and developed a new field of nano-separation sciences.

457175 Immobilization and Orientation Controll of Anti-CRP Scfv on a Seuface of Polystyrene Latex Beads and Its Application to Latex-Turbidimetoric Assay

Abstract:

In this study, we investigated immobilization of anti-C-Peactive Protein (CRP) single-chain Fv antibodies onto a surface of polystyrene latex beads for use in latex turbidimetoric assay of CRP. Both Polystyrene-binding peptide (PS-tag) and Poly(methyl methacrylate)-binding peptide-(PMMA-tag) were introduced to its C-terminal region for site-specific attachment and orientation control of scFv, while scFv with D10-tag which was decapeptide only composed of Asp was used as a negative control. Three types of scFvs were over-expressed in an insoluble fraction of E. coli cell, and highly purified via IMAC under the reduced and denatured condition. Conformation of tagged scFvs were refolded by one-step dialysis using 50mM Tris-HCl, pH 8.5 for PMMA-tagged scFv (scFv-PM) and D10-tagged scFv (scFv-D10), and 50mM MES, pH 6.0 for PS-tagged scFv (scFv-PS). Consequently, according to the results of indirect ELISAs and sandwich ELISA, antigen-binding activities of all scFvs tested were fairly recovered by the refolding operation.

ScFv-immobilized PS latex beads were prepared by the passive adsorption method. Averaged diameter of scFv-PS were significantly increased probably because multi-point interaction of positively-charged PS-tag induced aggregation of PS latex beads which were negatively-charged. On the other hand, latex beads with scFv-PM and scFv-D10 were well-dispersed at the pH between 7.5 and 8.5, while scFv-PM-immobilized latex beads were aggregated at pH 6.5, suggesting that surface charge of scFv-PM-immobilized latex beads might be neutral at this pH which was correspondence to isoelectric point of scFv without PMMA-tag. These results expected that scFv molecules were site-specifically immobilized through the PMMA-tag. ScFv-D10-immobilized latex beads were not aggregated at the pH range between 6.5 and 8.5, suggesting that scFv-D10 molecules were randomly oriented and change in charge density of latex beads was not significant so much because the exposed D10-tag might be caught up on surface charge density which must be reduced by immobilization of scFv.

Turbidimetoric assays were performed using both scFv-PM and scFv-D10-immobilized latex beads. Consequently, scFv-PM-immobilized latex beads had wider detection range of CRP, compared with scFv-D10-immobilized latex beads.

Thus, scFv-PM-immobilized latex beads developed in the present study will be useful for economic and sensitive detection of biomarkers using latex-turbidimetric assay.

466400 Enzymatic Conjugation Strategy for the Design of Artificial Biomolecular Assemblies

Abstract:

A variety of functional biomolecular assemblies are observed in natural biological systems. In particular, proteinaceous supramolecular assemblies, such as viral capsids and bacterial surface pili, are of great interest in bioscience and bionanotechnology because their well-organized structures associated with unique biological functions would give us insight in the design of artificial biomolecular assemblies in practical applications. To mimic natural biological assembly systems, we have established site-specific protein conjugation techniques by using two enzymes with different catalytic properties.

Microbial transglutaminase (MTG) that works in crosslinking reaction between the side chains of glutamine (Gln) and lysine (Lys) residues gives us opportunity to attach an artificial affinity moiety to the selective site of protein of interest (POI). We have designed artificial substrates with multiple biotin moieties, and conducted site-specific biotinylation of POI. As a proof-of-concept study, tetrabiotinylated endoglucanase was prepared, and mixed with streptavidin. Eventually, one-dimensional assembly comprised of these protein building blocks was obtained, and it exhibited the unique hydrolytic activity for cellulosic substrates.

Another enzyme candidate is horseradish peroxidase (HRP) that catalyzes radical coupling reaction between the phenolic side chains of tyrosine (Tyr) residue. We found that HRPâ€™s substrate specificity can be narrowed to a peptide tag fused genetically to POI. In line with the HRP-catalyzed polyphenol formation, we succeeded in creating proteinaceous polymeric materials via the site-specific radical coupling at a Tyr-containing tag (Y-tag). It is worth noting that both MTG and HRP provide us with site-specific and covalent conjugation strategies for a range of biomolecules, which is useful for the design of artificial biomolecular assemblies.

462831 Optimization and Scale-up of Microfiltration TFF for Reliable Clarification Processes

Abstract:

One of the most challenging steps in the manufacture of microbially produced therapeutics is the clarification of feed solutions containing solubilized inclusion bodies. In such applications the recovery and purity of the target protein is often limited. The high percentage of impurities results into low depth filter capacities applying the normal flow filtration (NFF) and insufficient separation efficiency in addition to the scale-up limits of the centrifugation. Tangential flow filtration (TFF) with a microfiltration membrane has been considered in such processes. Importantly, the control of permeate flow rate is essential to achieve filtration success criteria and limit fouling.
This study highlights the successes of a permeate-controlled TFF microfiltration process at various scales with different feed streams. The laboratory TFF system with a membrane area of 0.1 m2 was used for the initial evaluation. A pump controlled the permeate flux ensuring a stable and successful process operation. By using the model feed (BSA containing Whey feed stream), the protein sieving was investigated. The process simulation has confirmed the cake-mass resistance profile and protein sieving prediction. For the clarification process of solubilized IBs from Escherichia coli, the permeate-controlled TFF set-up was compared to the NFF process in regards to separation efficiency. Although the TFF membrane used has a smaller nominal pore size than the used depth filters, a Pellicon 1000 kDa PES membrane in V-screen cassette provided higher retention of impurities. The resistance curves were analyzed applying the cake-lift model for TFF and cake model for NFF. The overall fittings suggests that the advantage of the permeate-controlled TFF might be due to the control of the membrane pressure increase which results in low membrane fouling, effective separation and high protein sieving. The scale-up to the Phase I GMP runs demonstrated the robustness of this technology and justifiably used in a clarification process.

464171 Design Calculation Procedure for Flow-through Chromatography

Abstract:

Flow-through chromatography (FTC) is considered to be an efficient method for separating two components as the flow is continuous. It is also suited for a single-use chromatography operation. Although several papers have been published on the flow-through chromatography process design, it is still not clear especially how the mass transfer affects the separation performance.

In this study, we have demonstrated how to design FTC based on the distribution coefficient as a function of salt concentration $K(I)$ obtained from linear gradient elution experimental data along with the zone spreading (mass transfer) data. We have determined K-I curves of monomer and dimer from linear gradient elution (LGE) experiments. The zone spreading parameter values were evaluated based on the plate number N by using a simplified single-lumped zone spreading parameter model. Curves for monomer and dimer were numerically calculated in order to determine the sample feed volume VF that can allow complete separation. Then, VF values were calculated as a function of I. For low K values, the operation time was short whereas VF was quite small. By increasing K values (which corresponds to lowering I), VF increased. However, the separation time increased markedly. The separation time – the sample feed volume curves were found to be useful for designing FTC. It was also found that for difficult separations such as aggregate removal the mass transfer plays a significant role. Flow-velocity must be carefully tuned in order to increase the sample volume. For such cases monoliths or membranes are better choice.

464386 Purification of Therapeutic Proteins By Affinity Precipitation Using Peptide Conjugated Smart Biopolymers

Abstract:

While chromatographic bioprocessing is the dominant mode of separation for biological products, there is an ongoing search for more cost-effective and scalable processes to address the increasing diversity of therapeutic biomolecules and their increasing titers. Affinity precipitation using smart biopolymers has generated much interest recently for its potential for the simultaneous recovery and purification of biological products. The current work focuses on designing affinity peptides against therapeutic proteins followed by the recombinant engineering of elastin like polypeptide (ELP)-peptide conjugates to develop affinity precipitation processes for proteins. Peptides were designed using epitope mapping as well as high throughput approaches such as phage display. Promising leads were then examined using microarray screening and fluorescence polarization in order to establish peptides with sufficient binding affinity as well as elutability using appropriate modifications in the fluid phase conditions. ELP-peptide constructs against model proteins and proteins of therapeutic importance were recombinantly engineered and their efficacy for purifying the proteins by affinity precipitation were evaluated. Further, the effects of different factors on the precipitation process were studied, such as (i) peptide presentation on the ELP, (ii) peptide affinity towards target protein, (iii) absence and presence of amino acid linkers between the peptide and ELP and (iv) nature and length of these linkers. This study will help to develop design heuristics for ELP based constructs to facilitate effective protein purification by affinity precipitation.

461256 Adsorption and Separation of Native and Pegylated Proteins on Anion Exchange Resins with Varying Degrees of Polymer Grafting

Abstract:

Ion exchange resins containing charged polymer grafts have been shown to enhance protein binding capacity and mass transfer rates as a result of the dynamic nature of the polymer structure. Most of the previous work in this area has, however, been devoted to cation exchange resins. In this work, we have examined the adsorptive and chromatographic properties of native and PEGylated proteins on a range of anion exchange (AEX) resins that are based on the same macroporous backbone matrix, but that contain different amounts of grafted polymers. BSA, PEGylated BSA, and thyroglubulin are used as model proteins covering a broad range of molecular sizes and strength of interaction with the charged ligands on the resin. Adsorption isotherms and batch uptake kinetics were determined for each protein in order to quantify the macroscopic behavior, while confocal laser scanning microscopy (CLSM) was used to elucidate intraparticle transport for single and multicomponent adsorption. The results show that the performance of these resins depends on a balance of size-exclusion and electrostatic interactions. Polymer grafting increases the binding capacity and leads to an enhancement of adsorption rates for the smaller proteins. On the other hand, as molecular size increases a point is reached where size-exclusion becomes dominant resulting in poorer performance. An optimum grafted polymer content seems to exist that optimizes the balance of these two properties.

Introduction to the Fundamentals of Project Management

Paper # 474819
Dr. Eldon R. Larsen
Marshall University

In this paper we discuss several important concepts which will be helpful throughout the rest of the tutorial. First, we discuss the very important concept of *principles*. Then, we will seek to understand what we really mean when we say we have a project—in other words, "What is a project?" Finally, we will introduce the most important keys to successful project management, which will really be the foundation for much of the rest of the tutorial.

Principles: The Foundation

To begin, we need to understand the concept of what a *principle* is and how we can use that concept to strengthen our ability to manage projects. In our discussion of principles, the following definitions help us to understand some important ideas applicable to project management.

Tools. Tools are instruments, concepts, procedures, or techniques used to help us accomplish work, either physical or mental. A Gantt chart, to be discussed in a later paper, would be an example of a project management tool. Tools are not our masters; they are intended to be our helpers or servants. Tools can be applied in many ways and they can be misapplied in many ways. Tools need to be maintained properly in order for them to continue to help us effectively. We use tools in applications of rules, policies, and principles.

Rules. Rules are prescribed guides, procedures, or obligations for conduct or action and are sometimes binding by regulation or legal precept. An example project management rule might be that we don't have employees work overtime for more than two weeks straight without a two-day rest break before the next overtime period. Good rules are based upon some kind of principle(s), while bad rules probably have little or no foundation in principles. Rules occasionally need to be changed, especially as times and circumstances change. In our example overtime rule, a true emergency might require an exception.

Policies. Policies are very much like rules. In fact, they are often interchangeable in definition. They often consist of management procedures based on material interests. They might involve a definite course of action selected from among alternatives and in light of given conditions to guide and determine present and future decisions. An example policy might be that we will give an employee one day off if he or she needs to attend the funeral of an immediate family member. Like rules, policies do change and often need to change with time and with new circumstances. There might need to be exceptions to the example funeral policy, if the funeral was more than a given distance from the workplace. Like rules, good policies are based upon some kind of correct principle(s), while bad ones are not.

Theories. Theories tend to be derived from various analyses of facts and their relation to one another. They usually are plausible or scientifically acceptable general concepts offered to explain phenomena. They usually are based on observations and are: hypotheses assumed for the sake of argument or investigation; suppositions or conjectures; or abstract thoughts or speculation. Because theories are not facts, but rather are attempts to explain facts or observations, theories often change or need to change, especially when new facts are obtained or when new observations occur. It is not uncommon to read in the newspaper of some theory that has changed due to new information which someone has obtained. In project management, we occasionally encounter various management theories. Some are well-based on extensive evidence, while others are less-well based.

Values. Values are concepts or perceptions which we highly rate or highly regard—something we prize. They may or may not be correct perceptions, but to each of us they are important in our decisions and actions. Religions, philosophies, personal ideals, organizational ideals, and some morals might fall under the category of values. For example, it might be that a person highly values self-reliance to the point that he or she feels it is important to become wealthy before truly being self-reliant. His or her value is that he or she needs to become wealthy. However, it is not unusual for people and organizations to change their values over time, even if only a little, as a result of experience, knowledge, and choices they make in life. In our example value of the need to become wealthy, the same person may encounter life in such a way to cause him or her to re-evaluate the desire to become wealthy and replace it with a desire to sacrifice of higher causes. Thus, one's values do occasionally change.

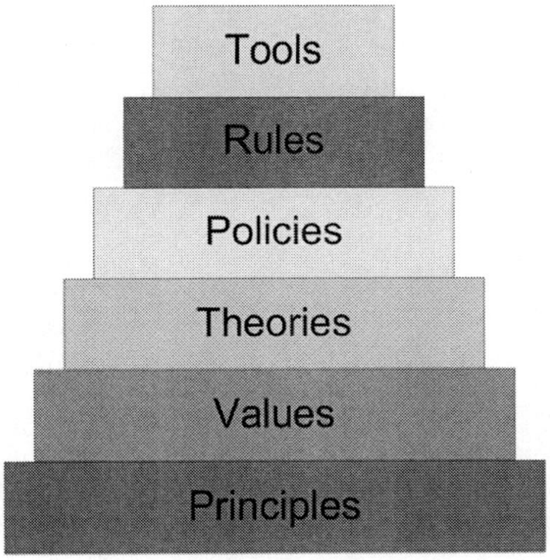

FIGURE 1Good Project Management Should Be Founded Upon Principles

Principles. Principles are fundamental or foundational unchanging truths or natural laws that cannot be broken. Principles are different from all the other concepts and definitions above in that *they do not change* with time or circumstance and they are true for everyone; regardless of background, religion, or philosophy; regardless of whether we choose to ignore them. They are comprehensive and they are self-evident. A few examples of principles are given in Table 1.

Principles are not the same as the *application* of principles. Applications of principles might change a little or a great deal as time or circumstances change.

For example, it is important that we love people—this is a principle. In some cases this might mean giving something to someone in need, like giving food to a child who cannot provide for himself. In other cases, loving an adult might be applied by *not* giving him something he needs, particularly if it is important for that adult to learn how to obtain that thing through his own skills and initiative, so that he can grow and become more self-reliant—note that the importance of growth is also a principle. In the examples of the child and of the adult, love is evident, but the practice or applications of the principle of love are very different.

Table 1: Examples of Principles

The need to love and care about other people
The necessity of integrity
The importance of honesty
The importance of service to other people
The importance of human bonding
The need for patience
The need for good communication
The need to be loved
The need for individual growth
The need for harmony with one's conscience

We must note that while we can try to ignore principles, they never go away, they are always there. We can try to say that they are not true, but that does not change that they are true; and when we try to break principles, we must suffer the natural consequences, like we do when we try to break natural laws.

Because principles are so very fundamental, unchanging, and universally applicable, it is vital that we learn them, understand them, and learn to apply them correctly. This is true in project management and in any other aspect of life.

Throughout the use of this text, you are encouraged to focus special attention on learning the *principles* of project management. Although we will be considering a number of procedures, tools, and policies or rules, all these can all change over time—they will change as a function of circumstances. The principles however *will not* change.

If the reader deeply understands the principles of project management, which will never change, then he or she will have a firm foundation upon which to build various different applications of those principles, according to the needs of the situation. Then, we can also understand how to make better adjustments or exceptions to rules and policies as needed.

What is a Project?

Before we go any farther, we really ought to understand what a project is. This might seem trivial, but it is not. In fact many individuals and teams have started out to accomplish a "project" only to find later they really did not have any clear idea of what was to be accomplished and when.

There are many people and organizations who have attempted to develop a definition for a *project*. Because it covers the basics of all projects, we will use the following definition: A project has two main characteristics: (1) it is temporary, or time

The bottom line is this: whatever the situation, there should be a clear expectation as to the time frame for the project to start and end, along with any tolerances in those start and end times.

Unique. By the time a project is completed, a that project should provide one or more unique products, services, or results—it should have accomplished one or more specific objectives. *Unique* does not mean that similar projects have not been done before, but rather that every project is unique in some way.

The project must have specific objectives to be accomplished. Without specific objectives, the project team's work will be inefficient and ineffective. The less specific the objectives are, the greater the likelihood that there will be much wasted effort, increased rework, and missed opportunities. In addition, if the objectives are not clear, the project stakeholders will have a difficult time agreeing on when the project has been completed and whether or not the project has provided the product, service, or result each stakeholder anticipated.

Non-project Work. Work that falls outside of project work is referred to as being *operational work* or *maintenance work*. These types of work tend to be ongoing and repetitive. At home for example, the work of washing the laundry, cooking the meals, washing the dishes, and cleaning the house is very repetitive and ongoing—it seems it never ends! Although people might say that cleaning the house is a project, for the purposes of project management we normally would not consider that type of work as *project work*, but rather we consider it as *operational work* or alternatively *maintenance work*, whichever you prefer—it is not unique and really is not temporary.

What is a Project Manager?

The obvious answer to the question, "What is a project manager?" is that a project manager is someone who manages projects. Perhaps we should re-phrase the question into the following: "What characteristics make up a good project manager?" Table 3 gives a list of characteristics that the author has compiled over the years which cover most of the aspects of what is needed to be a good project manager. The list is large and yet probably lacking. Good project managers need to have many talents and many excellent character traits if they are to be successful project managers and project leaders. Many of the characteristics in Table 3 are covered in detail later in this text.

Table 3: Characteristics of a Good Project Manager

Maintains high integrity
Excellent communicator and listener
Trustworthy
Honest and ethical
Friendly and courteous
Respectful *and* respectable
Organized
Teacher and builder of people
Patient
Accountable and asks for accountability
Cares about people
Flexible when needed; not flexible when shouldn't be
Motivator
Professional in interactions with others
Good decision maker
Planner, with other stakeholders
Good manager of time
Good record keeper
Has good technical and managerial credibility
Familiar with the organization's culture, policies, and procedures
Good diplomatic skills
Receptive to suggestions
Skilled in developing team solutions and facilitates group participation
Doesn't ask others to do what he or she is not willing to do
Leads rather than pushes
Values the differences in others and their perspectives
Seeks synergy and win/win solutions
Helps others set goals in support of the project and organizational objectives

Keys to Successful Project Management: The Vital Three

Let's suppose that you are nearing the end of this project management text. You meet a friend in the elevator who says, "I understand you attended a tutorial about project management. What have you learned?" You have until you exit the elevator, about 20 seconds, to tell your friend what you have learned. What would you say? Well you don't know right now because you haven't completed this presentation. So, I am going to tell you what I hope you will say. You see, I am going to tell you right now what I consider to be the three most important concepts in this tutorial. In the elevator, you would likely only have time to tell your friend the three most important concepts anyway. By me telling you now what I want you to say, I am hoping that you will pay special attention to these items throughout the course of your study. The three most important keys to successful project management are given in Table 4.

Table 4: The Vital Three Keys to Successful Project Management

First*, the importance of* ***people*** *in project management!*

Second*, the importance of excellent definition of* ***project objectives****!*

Third*, the importance of excellent* ***planning*** *and* ***re-planning****!*

There are many concepts, processes, and principles important in project management and they are all important, but I consider these three to be the most important keys. If anyone of these is missing or weak, the entire project will suffer tremendously. Sometimes I refer to these as the *vital three* because they are so preeminently important. You can often succeed by neglecting some other aspects of project management, but you cannot ignore any one of the vital three—they are essential.

References

Covey, Steven R., *The Seven Habits of Highly Effective People.*

PMI. *A Guide to the Project Management Body of Knowledge (PMBOK® Guide)*, fourth edition, The Project Management Institute, Newtown Square, Pennsylvania, 2008.

The Importance of People in Project Management

Paper # 474820

Dr. Eldon R. Larsen
Marshall University

Of the many important aspects of project management, the most important key to successful project management is people. When we speak of project management, we often think of Gantt charts, time lines, budgets, resource constraints, quality challenges, performance measures, and so on. While these are important, they are neither meaningful nor effective without people. This paper presents and discusses several fundamental concepts regarding the importance of people in project management.

Introduction

In 1993 when I first became deeply involved with formally studying and applying the principles and tools of project management, I made an appointment to visit a man who was well respected at for his project management expertise. He was Charles Brod, the Chair of the UCC (Union Carbide Corporation) Project Management Council. This council was made up of the most experienced project managers in the corporation.

I had been asked by UCC to lead a team to develop and implement project management principles, techniques, and tools specifically for research and development, an area often not accustomed to the rigors of formal project management. I had been on many research and development project teams and product or process commercialization teams, once serving on twelve different teams at the same time. However, I had little formal project management training at the time. I knew I could use Charles's insights for this effort I had been assigned.

I did some preparation and wrote questions to ask Charles, so as to make the most of our visit. As I entered his office at the appointed time, and before we sat down, I said, "Charles, I know that people are important in project management..." Before I could finish my sentence he pointed his finger at me and said, "You stop right there! People aren't just important in project management, people are project management!" (Brod, 1993) I just stood there stunned for a few seconds as my brain rapidly processed what Charles had said: "People are project management!"

As my mind raced through my many experiences in the many projects I had worked on, I knew almost instantly that what Charles had just said was not only true, but was the verbalization of one of the most important principles of project management. In comparison to that statement, little else mattered in that interview.

Principle:

People are project management!

People Are Project Management

Jack Gido and James Clements (Gido and Clements, 1999) say it this way: "It is the people—not the procedures and techniques—that are critical to accomplishing the project objective. Procedures and techniques are merely tools to help people do their jobs."

As people progress in their careers, many become project managers. These people are often highly trained technically and tend to be very analytical—this helps them to be good technical problem solvers. However, technical professionals often have had little formal training which will enable them to effectively lead and manage people. Thus, they could be weak people-problem solvers. Yet, this is exactly what they are being expected to do as they move into project management or other management positions.

People in new project management positions might well ask themselves the following questions: "What do seasoned project managers wish they had known on their first project management assignment?" "What are the responsibilities?" "What are the people interactions that take place?" Following are some of the things new project managers would have liked to know, based on research by J. Pinto and O. Kharbanda (Pinto and Kharbanda, 1995).

- Realizing that conflicts occur as a natural part of team development,
- Knowing who are the project stakeholders and what are their agendas,
- Realizing that organizations are political,
- Understanding that project management is leader-intensive and you must be flexible,
- Realizing that customer satisfaction is essential for project success,
- Understanding that you must be a motivator, coach, peacemaker, and conflict manager,
- Understanding that your team develops attitudes based on your emotions,
- Knowing that you always need to ask "What if…"; don't be comfortable with the status quo, and
- Learning that above all you must plan, plan, plan!

All of the above learnings involve *people interactions*!

Project managers have responsibilities among other things to (1) define and clarify objectives, (2) lead and manage, (3) plan and re-plan, (4) organize, and (5) control and influence. Again, all of these responsibilities involve people interactions.

The following sections of this paper discuss and make suggestions for some of the multiple interactions among people involved with projects. The interactions discussed are

- Stakeholders and stakeholder management
- Project manager interacting with project team members,
- Project manager interacting with senior managers,
- Project manager interacting with line mangers,
- Project manager interacting with the other employees within his or her company,
- Project manager interacting with the general public outside of his or her company,
- Project manager and team interacting with the customer, and
- Project manager and team interacting with contractors and other internal or external service providers.

Stakeholders and Stakeholder Management

All projects have project stakeholders. Examples of potential stakeholders are the customer, the project sponsor, senior managers, the project manager, the project team, organizational line managers, contractors and subcontractors, service and trades personnel, landlords and/or leaseholders, suppliers, recipients of products and technology, government entities and regulators, and possibly the surrounding populace.

A wise project manager will do what is needed to identify all significant stakeholders. The project manager will recognize that some stakeholders that might not seem significant at the present, but they might become significant and so will need to be monitored. One major mistake many project managers make is not to do a careful analysis of all stakeholders.

Stakeholders can be either positive stakeholders or negative stakeholders. Generally, positive stakeholders will benefit from the project being a success and will want it to succeed. Negative stakeholders will likely not benefit from the project's success, and so will want it to fail. Some stakeholders that currently are positive might change and become negative, and vice versa.

You might need the full support and cooperation of some stakeholders in order for the project to succeed. Therefore, you should always ask yourself the following questions: "Who will benefit from this project? Who might be hurt by this project? Who might otherwise be impacted by this project? Who will be key in helping this project to succeed? Who else might I have forgotten about?"

Because of the many concerns, needs, and expectations of stakeholders, it is essential that the project manager conduct an effort to identify the stakeholders, their expectations, their potential impact on the project, how the project impacts them, what is needed from them, and so forth. To do this, the project manager develops a *Stakeholder Register*.

Table 2.1 shows an example of a possible stakeholder register. Here we identify the stakeholders, their positions in the internal or external organizations, their power and authority, their location, how to contact them, whether they are positive or negative stakeholders, their relationship to the project, what influences they might have upon and for the project, their needs—must have, their wants—like to have, and what concerns or fears they might have about the project. We also very clearly identify their requirements for the project and their expectations beyond their requirements. We clearly identify our strategies for managing their requirements and expectations so that changes are clearly analyzed, controlled, approved or not, and communicated. Finally, we clearly define communication needs for each stakeholder.

The rest of this paper discusses some specific needs and helps in dealing with specific types of stakeholders.

Project Manager Interacting with Project Team Members

Managers are not necessarily the same as leaders. A manager has assigned people that he or she is to manage. Leaders are different. There are no leaders without followers, and people choose to follow those they consider to be their leaders. Leadership and management are all about working with people.

Much like a music conductor (Gido and Clements, 1999) leads an orchestra to play a concert, the project manager leads the project team to accomplish the project objectives. The project team, like an orchestra, must be unified and in tune (aligned), as they follow their leader. For there to be harmony, the team must want to follow the leader—this is a vital people interaction principle. The leader and team must care enough about each other to make this happen.

Project managers and the project team members interact in many ways. They do this as they define and clarify project objectives; plan and re-plan; make and accept task assignments; delegate and expect accountability; manage and resolve conflict; reward and discipline; generally communicate; give performance feedback; motivate and inspire; and develop personnel. Following are some suggestions for project managers to help them to achieve success in these interactions with team members.

- When practical, involve the project team in refining the project objectives with the customer;
- Assure that the project objectives, constraints, and deliverables are clear and that they are understood by all team members;
- Always involve the project team in project planning, then...
- Seek commitment from team members;
- When challenges arise, genuinely ask for input from team members;

Table 2-1 Example Stakeholder Register

STAKEHOLDER REGISTER					
Stakeholder Characteristics					
Stakeholder Name	Position	Power & Authority	Location	Contact Information	Category: Pos./Neg.

Stakeholder Relations and Needs, Wants, and Fears					
Stakeholder Name	Relation to the project	Potential Influences	Needs	Wants	Fears

Stakeholder Requirements, Expectations, Strategies, and Communications				
Stakeholder Name	Stakeholder requirements statements based on needs, wants, fears, and other information	Stakeholder additional expectations with respect to and for the project	Strategies and issues or concerns for managing the requirements and expectations	How, when, and what to communicate with this stakeholder

- Delegate as much as you can to team members, according to each individual's readiness to accept the added responsibility;
- Build and develop your followers; give them growth opportunities, but don't overwhelm them;
- Whenever feasible, minimize unnecessary bureaucracy;
- Learn what motivates and inspires each team member;
- Ask for accountability, while helping team members to get the authority and other resources they need in order to be able to be accountable;
- Work with line management to assure that team members have up-to-date tools;
- Reward or discipline team members sooner, rather than later;
- Make project meetings and communications valuable—don't waste people's time; (Larsen, 2001)
- When conflict arises, manage it quickly and seek win/win resolutions;
- Make a project communications plan and use it; and
- Keep accurate and up-to-date project records—this helps to minimize miscommunications and unnecessary rework.

Project Manager Interacting with Senior Managers

There are many reasons that a project manager needs to interact with senior managers and to nurture these relationships. Senior managers

- Assure business goals are being achieved,
- Often act as the interface between the customer and key project personnel,
- Provide essential support and guidance for projects, and
- Often act as project sponsors or champions.

It is absolutely essential that effective interactions occur between the project manager and senior managers. If this interaction with senior management is not effective, support for the project could wane, misunderstandings will almost surely occur, project objectives and priorities will either be unclear or insufficiently acted upon, and the project team could easily feel itself drifting without a good sense of direction or support. Some of the interactions that often occur between the project manager and senior managers have the following purposes.

- Define project objectives and deliverables,
- Clarify and align business and technical objectives,
- Clarify company policies and procedures,
- Clarify project scope and project risks,
- Clarify critical success factors,

- Clarify key project priorities or drivers and tolerances in the triple constraints of time, resources, and performance, (Larsen, 2002) and
- Finalize and approve the project charter.

In order for the above interactions between the project manager and senior managers to be effective, consider the following suggestions.

- Be honest, maintain integrity, be respectful;
- Don't be afraid to ask for or give clarification or more information as often as is needed;
- Give and ask for the accurate, whole picture, not just the good news;
- Be concise in communications, but be clear;
- Do all you can to prevent surprises;
- Be sure to thoroughly discuss and understand all important project risks, tolerances, and priorities;
- When discussing concerns, focus on issues rather than people;
- Wisely use senior management to help with conflict and risk management when their help is truly essential; and
- Never backbite your senior managers—it won't help and will likely make things worse.

Project Manager Interacting with Line Managers

Line managers frequently play a vital role in contributing to project success or failure. In many organizations it is the line managers who select whom in their groups to assign to a given project. In many organizations it is also the line manager who actually pays the employees' salaries on a team. This can cause a definite allegiance preference for the line manager over the project manager when it comes to deciding which boss to follow when getting mixed direction signals. In organizations where line managers appoint team members, these managers

- Usually provide the people to work on the project;
- Assure that their people have proper training;
- Monitor the workload of their people;
- Often take care of the administrative needs of their people (payroll, safety certification, annual reviews); and
- Are usually much more technically expert than is the project manager.

From the above list it should be obvious that line managers can be vital to your success! Some suggestions for working with line managers are

- In the very early stages of the project, begin communicating with the line managers of people who might be selected to be on the project team;
- Assure that the project objectives, deliverables, priorities, drivers, and tolerances are clear to each line manger who will be providing resources or services;
- Develop and nurture a close working relationship—you need line managers' support, advice, wisdom, and buy-in;
- Always keep line managers informed of the project progress, concerns, challenges, and needs;
- Keep line managers aware of personnel achievements, performance issues, or needs pertinent to their areas of responsibilities;
- Listen sincerely to the concerns and advice of line managers—if you don't, you will pay for it; and above all
- Communicate well and regularly. (Larsen, 2003)

Project Manager Interacting with the Company and General Public

Depending on the project, a project manager's interaction with the company as a whole or with the general public can range from non-existent to frequent and crucial. Corporate or public perception of you and your project can have the potential to make or break your project—if this applies to you, nurture appropriate interpersonal relations with all the care you can give. Following are suggestions.

- Learn what major concerns the company or the public is likely to have and, as appropriate, be fully prepared to address them honestly and openly. If you can't be open because of secrecy agreements or other legal constraints, make that clear to those who are concerned.
- Always be prepared! Know exactly what is happening on the project. Then plan and practice how to communicate with those concerned. Make sure senior management, the customer, and the project team know what you will present and that they have given you their suggestions and support.
- Try to put yourself in the shoes of the people with whom you need to communicate. Seek to understand them (Covey, 1989).
- Always be ethical in your behavior and your words.
- Do not lose your cool! It takes many good interactions with people to make up for one bad one.

Project Manager Interacting with the Customer

The interaction the project manager has with the customer is perhaps the most important interaction of all, and could have the greatest impact on customer satisfaction—the real measure of project success. The customer, better than all others, usually knows most what he or she wants at the end of the project. Customer perceptions play a strong project-success role. Effective people-interaction is necessary to transfer customer expectations to the project manager and team.

Following are suggestions to help in this interaction.

- Unless the nature of the project prevents this, the project manager and team need close interaction with the customer as early in the project as possible—this is when exists the greatest potential to affect the final project outcome. Make this happen if possible.
- Maintain open channels of communication throughout the entire project.
- Early interactions between the project manager, team, and customer should include developing a very clear understanding of the
  - ✓ Project objectives,
  - ✓ Project priorities, triple constraint drivers, and tolerances,
  - ✓ Project's critical success factors,
  - ✓ Project's scope and deliverables,
  - ✓ Project background and justification, and
  - ✓ Any key concerns and assumptions.

Project Manager Interacting with Contractors and Other Internal or External Service Providers

Outsourcing of project work has increased greatly over the past decade. Flannes and Levin (2001) point out that those temporary workers sometimes "contribute a greater depth and breadth of experience." It might even be that the project manager is outsourced. The presence of contractors and other service providers, either part of the project team or not, presents important people-interaction challenges. Some suggestions follow.

- Remember, just as with the other key people interactions we have discussed, constant and nurtured communication is absolutely essential.
- Work early to address any cultural, technological, or other barriers that might impede this needed communication—make a plan to do this.
- Don't treat people not on the project team as second-class citizens. Value them. Care about them. Seek to understand their perspective. Your efforts will pay big dividends! Many of them could be project stakeholders.
- When dealing with contractors or people not directly on the project team, record-keeping and communication become even more important to avoid misunderstandings, re-work, and unwanted problems later on. Make sure important communications are in writing, and then follow up personally to discuss any questions or make clarifications. If changes are needed, put them in writing and get necessary approvals.

- Read all contracts thoroughly and make sure you understand them.
- If the other party is writing the contract, make a quick draft yourself of the contract before you see theirs. This will help you to have thought through what to look for and what questions to ask when you review their contract. After it is too late, many people wish they had more fully realized what they were signing!
- If you are writing the contract, ask the other party what they would like to see in it. Get input from other people who will be working on the project and people who have experience with contracts. Then, have a legal expert help you do the final contract to avoid problems you might miss. Make sure senior management signs-off on whatever the contract says. If you all work together, you are more likely to achieve win/win results.

People and Project Success

Notice how vital are the importance of people interactions and the perceptions of people, in addition to technical performance, as expressed in the following definition of project success from a study of approximately 650 projects (Baker, Murphy, and Fisher; 1974).

> "If the project meets the technical performance specifications and/or mission to be performed, and if there is a high level of satisfaction concerning the project outcome among key people in the parent organization, key people in the client organization, key people on the project team, and key users or clientele of the project effort, the project is considered an overall success."

This definition of project success places a large emphasis on perception—a people feeling. Notice also that schedule and budget are not explicitly stated as part of project success, although they undoubtedly somewhat affect perception and feeling. Nonetheless, such a definition certainly exemplifies the great importance of people in measuring and accomplishing successful projects.

Summary and conclusions

Good people interactions are vital to successful project management and must be carefully nurtured. These interactions permeate essentially every aspect of the project and are also the most time-consuming responsibility of the project manager. The above discussion should clearly demonstrate to the reader that people are indeed the most important key to successful project management. Because of the importance of people, a project manager needs to learn to sincerely care about each person associated with the project, and nurture these people interactions to achieve success. Charles Brod summarized it well, "People are project management!"

References

Baker, Bruce N., David C. Murphy, and Dalmar Fisher, "Factors Affecting Project Success," Ch. 35 of David I. Cleland and William R. King (eds.), Project Management Handbook, 2nd Ed., Van Nostrand Reinhold, New York, 1988.

Brod, Charles, Personal communication with the author, 1993.

Covey, Stephen R., The Seven Habits of Highly Effective People—Restoring the Character Ethic, Simon & Schuster, New York, 1989.

Flannes, Steven W. and Ginger Levin, People Skills for Project Managers, Management Concepts, Vienna, Virginia, 2001.

Gido, Jack and James P. Clements, Successful Project Management, South-Western College Publishing, Cincinnati, Ohio, 1999.

Larsen, Eldon R., "Not Another Meeting!—Planning and Conducting Effective Meetings," Proceedings of the ASEM National Meeting, October 13, 2001.

Larsen, Eldon R., "Start Projects Right—How to Develop Effective Project Objectives," Proceedings of the ASEM National Meeting, October 4, 2002.

Larsen, Eldon R., "Communicating With the Boss and Other Public Things," Management Topical: The Engineer in Transition to Management—Tutorial, Published in the Proceedings of the topical at the AIChE National Meeting, March 31-April 3, 2003.

Pinto, J. and O. Kharbanda, "Lessons for an Accidental Profession," Business Horizons, March-April 1995 (quoted by Gido and Clements, p. 83; 1999, cited in this paper).

Communication—A Better Understanding

Paper # 474822
Dr. Eldon R. Larsen
Marshall University

Today's business organizations often seem less caring, more informal, and having less long-term mutual-commitment. Our interactions are ever-increasingly more digital. Hence, the need for excellent and clear communications between employees, managers, peers, customers, and the outside world will continually become more vital to long-term successful enterprises. This paper presents principles, concepts, and tools to help the reader gain a new vision of communication and how to do it better in today's challenging environments.

Additionally, this paper presents ideas on what is truly needed for good communication—what foundation concepts are necessary if we really hope to successfully communicate. The paper discusses how communication takes place and reviews the very important concept of *empathic listening* as defined by best-selling author Stephen Covey (2004). Finally, it is important to make a communications plan when we need to manage communication.

Introduction

Communication is one of the most important concepts and tools in human life. It is also one of the most used and abused words in human interactions. Its real meaning is rarely fully understood. Significantly, when one really understands the meaning of communication, a paradigm shift often occurs. With this paradigm shift, we start to look at our interactions with other people in a whole new, deeply meaningful way. We open the door to truly effective human interaction.

We use communication in essentially every interaction we have with other people, be it verbal, written, silent symbolism, body language, pictures, sounds, or other ways. Yet, it is also communication, miscommunication, poor communication, or lack of communication which frequently gets us into challenging situations and can cause frustrating emotions, at least when we communicate in the way that people commonly define communication.

In interviews with hundreds of project managers which my graduate students have conducted over more than a decade, many of these project managers state that some of their greatest challenges come from working and communicating with people. Why do we have so much trouble with communication? Why is it such a challenge? What are we missing?

Someone once said, "I know that you believe you understand what you think I said, but I am not sure you realize that what you heard is not what I meant." (Dinsmore, 1990) Does this sound familiar? If so, you are not alone. It happens all the time.

This paper presents important ideas and concepts that can help. One of the most important concepts discussed is what the word *communication* really means.

A better understanding of the root meaning of *communication* can cause a very important paradigm shift, to the point that the word takes on a whole new meaning. It can give us new vision. This help opens the door to better communication for us all.

Do We Really Need Good Communication?

Project managers spend the vast amount of their time communicating in some way or another. They are constantly providing direction, asking for information, clarifying objectives, seeking for understanding, and giving feedback.

Whether you are an employee or a new or seasoned manager, it is important for you to realize that (1) bosses want to be kept informed, (2) bosses don't like surprises, and (3) bosses depend on communication to help them to plan and to organize, motivate, and control or influence people—actually the same could be said for most of us, not just bosses. Good communication is like the lifeblood of a successful project manager or a successful leader.

Project managers, team members, stakeholders, and others all struggle when we use poor communication techniques. Communication in its many-varied facets consistently comes up as a major managerial issue.

It is commonly stated that the three most important factors to consider in real estate are (1) location, (2) location, and (3) location. In management, especially in project management, the three most important factors to consider are (1) communication, (2) communication, and (3) communication.

What Is Communication?

What does the word *communication* really mean? Why ask this question about a word so commonly used? Everyone knows what we mean by communication—don't they? *Do we*? Let's look a little deeper.

High school English teachers often tell their students to look at the "roots" of a word in order to really understand it. These teachers often split words into various prefixes, suffixes, and main-word fragments, and then look at the Latin or other linguistic meanings of these word fragments to determine the true meaning of words.

Let's do this root-understanding-exercise for the word *communication*. We will look at the roots in French, rather than Latin, because the roots form common French words. *Communication* is in fact a French word derived from Latin, and is spelled the same in both French and English.

Comm/un/ication

First, we break communication into three word fragments: *comm, un,* and *ication*. *Comm* comes from *comme*, which is a separate French word that means "Like" or "As." *Un* in French means "One." Finally, *ication* essentially means "The process of becoming." When we put this all together the fundamental meaning of the word comes to light:

Communication = The process of becoming "like one" or "as one."

Let's say it again: Communication means the process of becoming like one or as one! It is the process of becoming unified!

Do we realize what this is saying? Think about the implications of the root-word definition! Communication is not just sharing or transmitting information. Communication is not just sending out a message. Communication is not just giving instructions. It is not just telling someone what you think or know. In fact, in and of themselves none of these are true communication according to the root-word definition. They can't be; they are one-way.

By the root-word definition above, communication almost always must be at least two-way. How else can we become like one? To be effective, this two-way interaction often needs to be very interactive and iterative. The only exception is the rare case when the receiver of the information just happens to fully agree with the sender of the information. Even then how do we know that communication has been achieved, unless we both interact and *communicate*?

Note that "The process of becoming like one" does NOT mean becoming the "same." We can be unified without being the same. In fact, we will be stronger and can synergize more effectively if we value differences, and yet are unified. It is a bit like having two pieces of a two-piece puzzle. The two pieces are not the same, but when they fit together and become *like one*, then they make a clear and complete, unified whole picture. Separately they are greatly lacking. We need to seek to become unified for project success.

In summary, true communication is a process of coming-together, in common understanding, vision, feeling and purpose, so that we truly understand each other to the point that we are becoming like one.

Principle:

Communication = The process of becoming like one

For most people, this concept, this original root-word definition, or could we even call it a *principle*, is a paradigm shift. It is a whole new way to look at what we are really trying to do when we say we want to communicate. It changes the whole picture or vision of what we are trying to accomplish. It changes our whole approach. The objective is not to just shove our opinions down someone else's throat! *We are trying to become like one.*

As a side note: If you physically try to shove something down someone's throat, you will likely get a response like gagging or vomiting. Think about that next time you try shoving your opinion down someone else's throat—it simply won't be pretty! It won't work, no matter how well intentioned you might be. This is not the way to communicate.

With this new vision—that we are seeking to become unified—we are now ready to realize that our very motivation in communication must change. Communication is not a one sided paradigm; it is two-sided at least, and it is serious and amazing. It involves us in a whole new perspective, not just our own perspective, and for a whole new purpose. How will we start communication anew with this new paradigm? We now are ready to investigate this more.

What Is Needed to Truly Communicate?

On what foundation are you to build if you want to truly communicate, if you want to become like one or as one with someone else? There are two main foundation building blocks I will discuss here.

1. Honestly and Mutually Care About Each Other

First, to truly communicate, you, and those with whom you are attempting to communicate, need to honestly and mutually care about each other. You must mutually respect each other and be considerate of each other's needs and perspectives. Without this mutual caring, communication efforts will quickly deteriorate to our unfortunately more modern one-way transfer of information we incorrectly call communication.

This presents a new challenge: Since you can't control the other person with whom you are trying to communicate, can this idealistic *communication* really happen? Isn't this a lost cause?

The answer is that you must start with yourself—only you can decide that you will really care about the other people around you, not just for your own self-interest, but for theirs too.

When you start with yourself, when you change your own motivations, almost miraculous interactions start to happen. For some strange reason, others seem to start to care about you too! They want more to work with you and to desire the oneness in purpose and vision and understanding that you emanate and that you too desire.

Thus, communication becomes a very personal issue. It is not an issue of better technology. "Communication among people depends not on technology, but rather on forces in people and their surroundings. It is a process that occurs within people." (Gibson, et al., 1997)

2. Mutually Align

Second, to truly communicate you also need to mutually align. Alignment means that you and those with whom you work all get pointed in the same direction, so to speak. You work toward the same goals, see the same vision and have a desire to work together toward achieving that vision. See Figure 1. One big question is "How do we accomplish this alignment of people?"

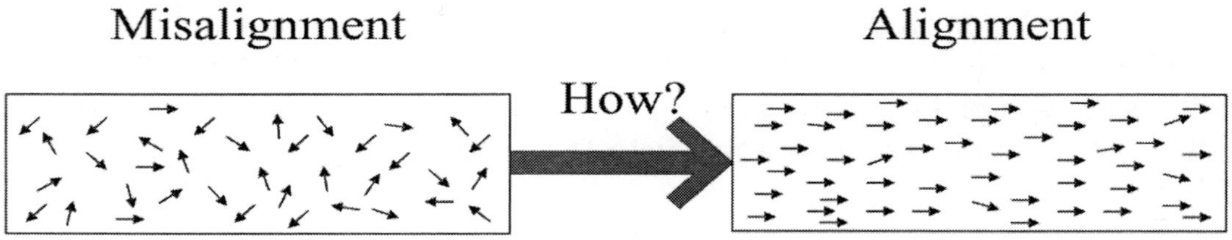

FIGURE 1 Transforming People from Misalignment to High Alignment

Let's look at an analogy to better understand personal alignment needed for better communication. When we take an automobile in for an alignment, Figure 2, what we mean is that we want to get all the wheels pointing in the same direction within certain tolerances. If the wheels are in poor alignment, the consequences are increased tire wear, steering difficulty, lower fuel efficiency, and extra wear on other parts of the car. Good wheel alignment is very important!

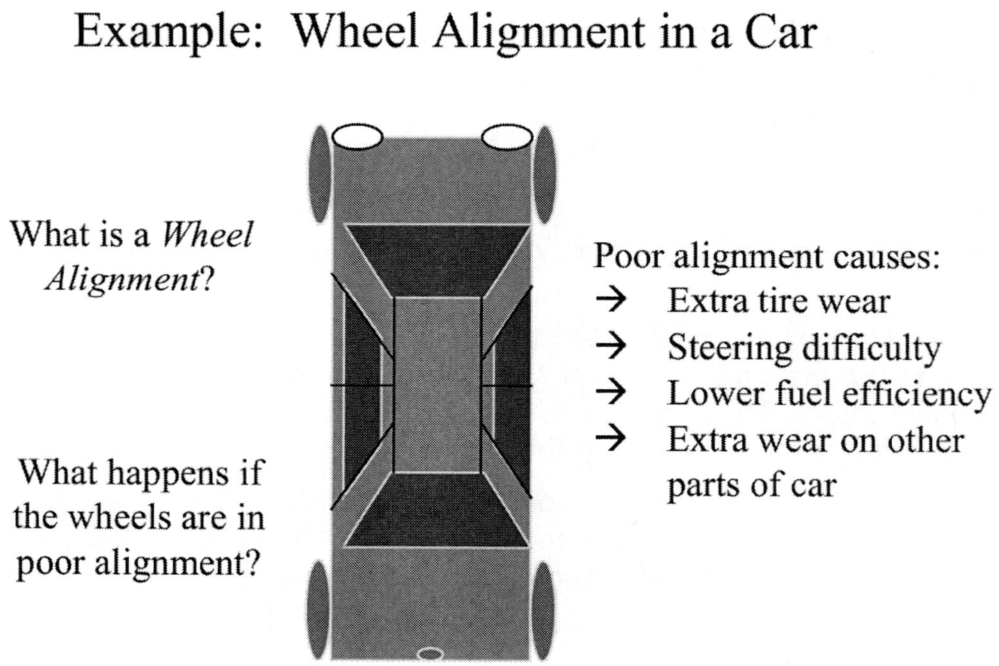

FIGURE 2 Good Alignment is Important for Cars—It is for People, Too!

Alignment of team members toward project objectives is equally important. Like the wheel-alignment analogy, poor team member alignment can lead to extra wear on the team members, difficulty to steer or manage the team, poor work efficiency, and even eventual wear on the larger organization. In contrast, a well-aligned team runs well, is much easier to manage, effectively and efficiently carries out the team's work, and improves the well-being of the organization as a whole.

Another analogy gives some insight into various ways we might try to achieve alignment. In this case, Figure 3, we use an iron bar that is not magnetized. In this bar the iron atoms are all spinning with dipole moments pointing randomly in an infinite number of directions. If we were to align enough of these iron atoms' dipoles, by use of some form of work, we could magnetize the iron bar.

This magnetization could be done by electrical induction, where we coil an insulated electric wire around the bar and pass a direct current through the wire. It could also be done by placing a permanent magnet next to the iron bar, with the work to align the previously random iron atoms coming from the permanent magnet. A third way to magnetize the iron bar would be to position the bar in a north-south orientation and pound on it with a hammer.

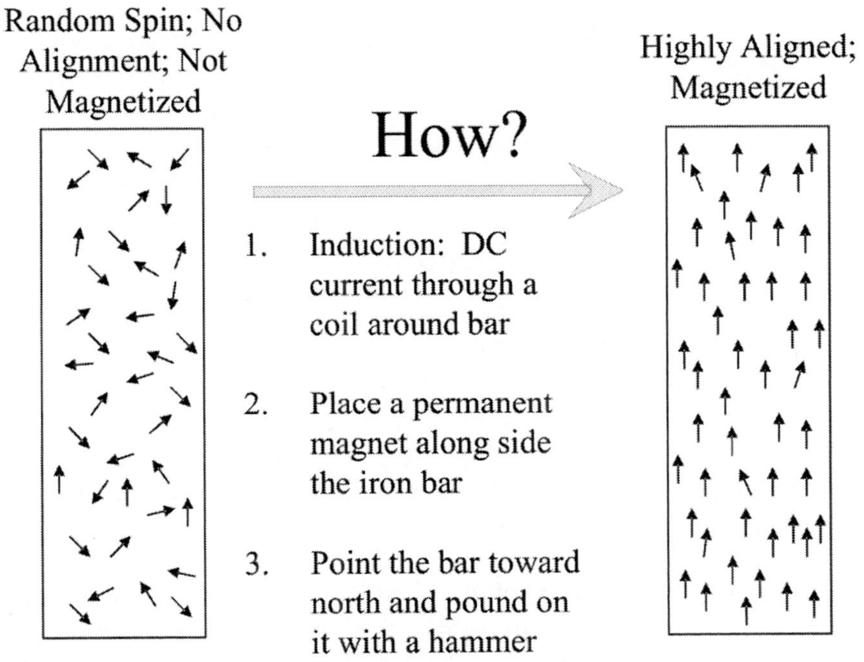

FIGURE 3 Alignment of Atoms in an Iron Bar

A group or team of people might be analogously aligned like the atoms in the iron bar. The team analogy to the iron bar is shown in Figure 4. Again, there are several ways that we might try to enhance team member alignment.

Similar to induction, we could put the team under the influence of some good interpersonal training and team-building exercises. Under this influence, the team is *induced* to start making better efforts to work together and to better understand each other.

Alternatively, we might expose the team to the example of people who have learned to permanently align themselves in order to achieve group objectives. These example-emanating people could show our team how to align by their good examples, or they might even become members of our team or work alongside us as mentors or partners. By their example and leadership, we learn how better to communicate and unify.

A less desirable approach would be to have the boss come in, point us in the right direction, and then "hammer" us into alignment by, for example, threatening us with dire consequences if we don't align.

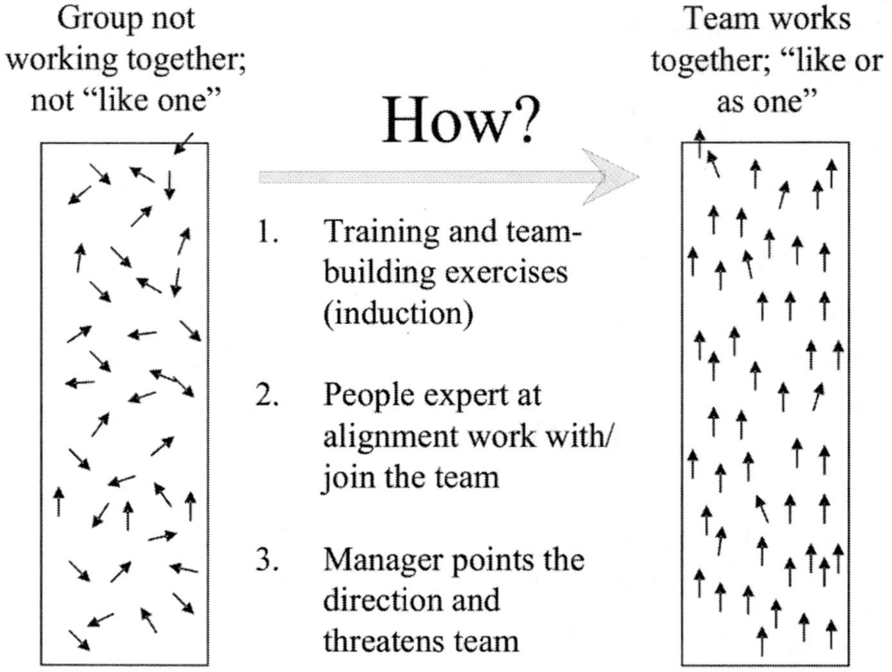

FIGURE 4 A Group of People Aligns into a *Team*

Must a group or team be perfectly aligned? No. In fact, a small degree of misalignment can have a few important benefits: (1) different perspectives, and the raising of concerns and alternatives, often aid our understanding, (2) creativity can be enhanced as we learn to accept and value differing perspectives, (3) strength can come from learning to manage minor personal-interaction challenges, and (4) people are more likely to feel they are in an environment of acceptance and freedom to communicate openly when their differences of opinion are valued.

There is a fifth benefit to having some lack of alignment: you might find, as you try to communicate with those who are not aligned, that there are some very important reasons those people are not aligned. For example, they might know something you don't, but they have been hesitant to share their knowledge. Through your sincere, caring communication efforts, you might very well find that the majority of the team is not pointed in the correct direction and that the few misaligned people actually were headed where the whole team really needs to be going!

On the other hand,, too much misalignment can cause (1) increased dissention, (2) decreased team efficiency, (3) decreased team effectiveness, (4) more arguments and backbiting, (5) lack of team spirit, and (6) increased chance of destructive behavior, and perhaps even attempts to sabotage the project. All of these problems are signs we are not really communicating—we are not becoming "like one." Therefore, high levels of misalignment in teams or individuals within a team must be dealt with, hopefully in a way that achieves better communication. Good use of the principles and concepts in this paper will help. If, after sincere application of these concepts, the matter cannot be rectified, *then* we might need to look at more drastic actions of removing a person from the team; but this should be a last resort, as this can have many negative consequences.

How Does Communication Take Place?

There are various modes and mediums through which we communicate. Examples are written words, speaking, body language, face-to-face meetings, telephone, voicemail, electronic mail, letters, memos, video-conferencing, and groupware. Of the above, *speaking* is one of the most common and important. Speaking is also a mode of communication where we often have difficulty. For example (Dinsmore, 1990), do you understand what former presidential aid Ron Ziegler meant in this statement:

> *You should not interpret by my use of "least likely" that ultimately, or when the final decision is made, that that may not be the decision, but what I am saying is that it is only one of the matters under consideration and the decision has not been made.* (Rees, 1979)

The above quotation made sense to the speaker, but is a little difficult for most readers or listeners to understand. How can we do better? Is there some model that will help us to understand better the process by which communication takes place? If so, it might give us an idea of how to improve communication.

The Effective Communication Model

One communication model that is found in many textbooks is called the Communication Process. (Gibson, et al., 1997) In this model there is a Sender who says some message in a certain way (medium) to a Receiver. The Receiver in turn gives feedback, indicating what effect the message had on the Receiver. The sending and feedback paths form a loop. This is shown in Figure 5.

Between various steps there are events involving encoding and decoding of the messages, which can muddle the message somewhat. These encoding and decoding events can be filtered or selectively influenced by personal perspective, bias, emotion, lack of understanding, errors, etc. Along the path there is also background noise caused by such things as disturbances, interruptions, side conversations, static, etc.

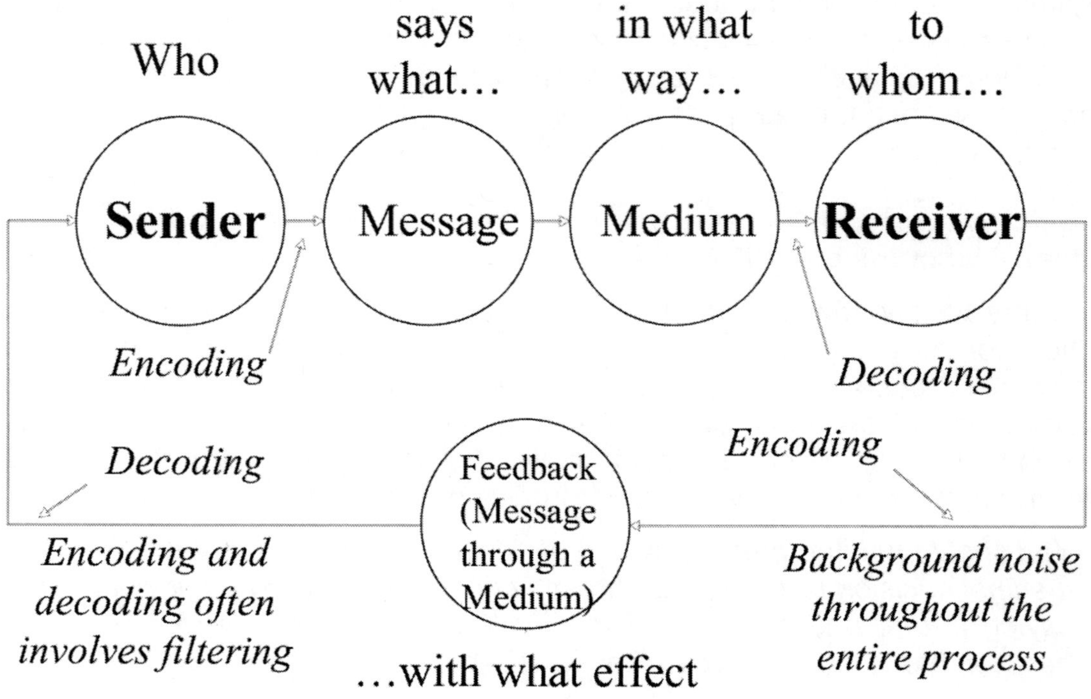

FIGURE 5 A Very Common Communication Process Model

At first glance the Communication Process model seems adequate, but in reality it tells only half the story. Communication is not just one way with some feedback—this is not how we usually become "like one" or "as one."

Instead, a more correct communication model shows that both of the people in the communication loop have information to share, both need feedback, and they must interact until they come to a common understanding—until they become like one. Figure 6 shows this improvement in the model, which I call the *Effective Communication Process Model*, which includes two-directional interaction. It is founded on principles of *mutual caring and alignment*.

To summarize, communication takes place through two-way interaction, with all participants having the opportunity to give their views as Senders and to give feedback as Receivers. In addition, effective communication must be founded upon mutual caring and alignment.

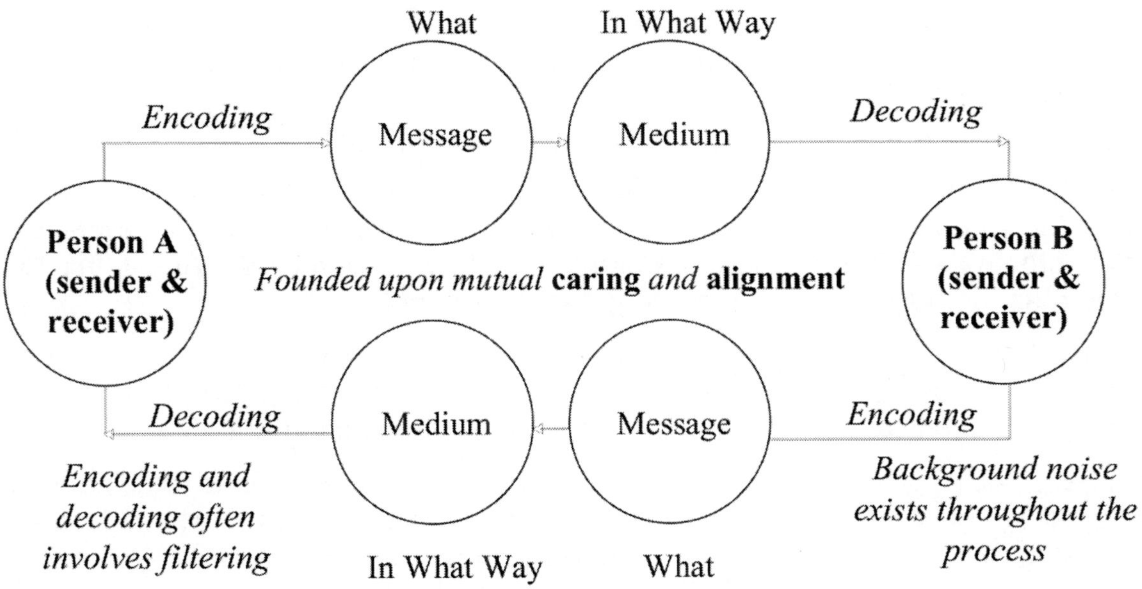

FIGURE 6 Two-Directional Effective Communication Process Model

Empathic Listening

Jack Gido and James P. Clements truly say, "The heart of communication is not words, but understanding. Not only to be understood, but also to understand. Half of making communication effective is listening." (Gido and Clements, 1999)

Much has been written about how to listen better, particularly on the topic of so-called Active Listening. But one of the most important listening concepts, when it comes to true communication as defined above, is the concept of *Empathic Listening*. (Covey, 2004)

What is Empathic Listening? Empathic Listening is not Active Listening, which is skill based. Instead, Empathic Listening is character and principle based. It is Active Listening along with sincerely caring about the other people with whom we are trying to communicate.

Empathic Listening = Active Listening + Sincere Caring

In Stephen Covey's bestselling book *The Seven Habits of Highly Effective People*, the fifth of the seven habits is

Seek first to understand, then to be understood." (Covey, 2004)

This habit is key to successful communication! We need to care so much about the other people with whom we are trying to communicate that we are willing to try to understand them, even before we try to have them understand us. We do this to increase *our* understanding, to get a different perspective, to learn new information, and to show the other person we truly care about him or her.

In addition, by understanding the other person, we greatly increase the possibility that we can synergize with them—to take the best of both of us and achieve things neither of us could have done by ourselves.

For many people this is a major behavioral change, but it's worth it! This change brings rewards far beyond most people's expectations—almost miraculous changes and rewards in some cases. Empathic Listening is probably the most important part of Covey's fifth habit—"Seek first to understand, then to be understood." Again, this form of listening must be built upon sincere caring for other people.

Covey has said, "If you use techniques or skills alone, I may sense duplicity, manipulation, or insincerity—I might wonder about your motives… I might feel over time that you really don't care so much about me as what you can get out of me."

Empathic Listening is based on truly caring about others. Someone once said, "If you really want to understand me, don't just listen to what I say, listen to what I feel."—This adds real meaning. Understanding other people, and then seeking to have them understand us, is vital to achieving the goal of becoming like one or as one.

To summarize, we need to Empathically Listen and seek first to understand, then to be understood.

Poor Listeners versus Good Listeners, and Miscellaneous Communication Barriers

Paul Dinsmore lists some typical characteristics of poor listeners and good listeners as follows: (Dinsmore, 1990). These are summarized in Table 3.1. Notice that you, the listener, are in control of most of the above good or bad listening habits!

Watch out, there are even more barriers to effective communication. For example, when trying to communicate, consider the following barriers in addition to those already listed above (some of the following are taking from Gibson, et al., 1997):

- Value judgments
- Semantics problems
- Jargon—technical, business, etc.
- Low credibility—low trust & confidence levels
- Proxemic behavior—physical distance between individuals (0-1.5 ft = intimate; 1.5-4 ft = personal; 4-12 = social; >12 feet = public)
- Assumptions
- Time pressures
- Differences in perception
- Emotions, either preceding or during communication
- Selective listening—daydreaming or preoccupations
- Thinking of counter-arguments while someone else is speaking
- The communication *medium*—telephone, email, etc.

- English as a second language
- Memory
- Distractions
- Too much formality
- Too big of group to properly interact to achieve unity

In spite of the fact that most of the above barriers are somewhat familiar to most of us, we still struggle with communications. Klaus Krippendorf says, "Despite the tremendous advances in communication and information technology, communication among people in organizations leaves much to be desired." (Gibson, et al., 1997)

Table 3.1: Poor versus Good Listeners (Dinsmore, 1999)

<u>The Poor Listener:</u>

- Always interrupts
- Jumps to conclusions
- Finishes my sentences
- Is inattentive; has wandering eyes and poor posture
- Changes the subject
- Writes everything down
- Doesn't give any response
- Is impatient
- Loses temper
- Fidgets with pen, pencil, or paper clip nervously

<u>The Good Listener</u>

- Looks at me while I'm speaking
- Questions me to clarify what I'm saying
- Shows concern by asking questions about my feelings
- Repeats some of the things I say
- Doesn't rush me
- Is poised and emotionally controlled
- Reacts responsibly with a nod of the head, a smile, or a frown
- Pays close attention
- Doesn't interrupt me
- Keeps on the subject until I've finished my thoughts

Plan to Communicate

In order to put effective communication into action, one additional tool can be very helpful: a communication plan. Someone once said, "There are three kinds of people in the world: those that make things happen, those that watch things happen, and those that wonder what happened." A plan helps you to "make things happen."

When making a communication plan, include (1) what to communicate, (2) with whom to communicate, (3) frequency and when to communicate, (4) format (memos or reports, for example), (5) how to communicate—the medium: formal or informal meeting, email, newsletter, voicemail, etc., and (6) who will do the particular communication effort. When making the plan, seek input from the key people who will likely be involved in the needed communication process. Their input can prove invaluable by helping you to understand their needs. A simple format for developing a communications plan is shown in Figure 7.

COMMUNICATIONS PLAN FOR PROJECT:______________					
What to Communicate?	To Whom?	Frequency?	Format?	How? (Medium)	Who Will Do It?

FIGURE 7 Example Format for a Project Communications Plan

Summary

In summary, remember: (1) good communication is critical to success, (2) communication is the process of becoming "like one" or "as one," (3) to communicate well, you have to truly care about other people enough to sincerely want to understand them, (4) "Seek first to understand, then to be understood"—empathically listen, and (5) make a communications plan and use it!

References

Covey, Stephen R., *The Seven Habits of Highly Effective People—Restoring the Character Ethic*, Simon & Schuster, New York, 2004.

Babcock, Daniel L. and Lucy C. Morse, *Managing Engineering and Technology—An Introduction to Management for Engineers*, 3rd ed., Prentice Hall, New Jersey, 2002.

Dinsmore, Paul C., *Human Factors in Project Management*, revised edition, AMACOM, American Management Association, New York, 1990.

Gibson, James L., John M. Ivancevich, and James H. Donnelly, Jr., *Organizations—Behavior, Structure, Processes*, Irwin/McGraw-Hill, Boston, Massachusetts, 1997.

Gido, Jack, and James P. Clements, *Successful Project Management*, South-Western College Publishing, Cincinnati, Ohio, 1999.

Mintzberg, Henry, *The Nature of Managerial Work*, Harper & Row, Publishers, Inc., New York, 1973, p. 39.

Rees, Nigel, ed., *Quote…Unquote*, St. Martin's Press, New York, 1979, p. 79.

Shainis, Murray J., Anton K. Dekom, and Charles R. McVinney, *Engineering Management—People and Projects*, Battelle Press, Columbus, Ohio, 1995, p. 33.

Planning and Conducting Effective Meetings

Paper # 474823
Dr. Eldon R. Larsen
Marshall University

Meetings can be viewed as necessary evils, valuable experiences, inspirational events, or intolerably boring curses. The very word *meeting* can cause various emotions to arise within us. This paper presents various rules and suggestions that lead to more effective and more enjoyable meetings. The approaches are founded on common sense and solid principles of management and human relations that are frequently overlooked. The rules and suggestions focus on reaching meeting objectives, effectively using agendas, nurturing good human relations, and ending meetings with an effective plan of action.

Introduction

Many people abhor meetings. Perhaps you have had similar feelings; many people do. Why do people dislike meetings? Common complaints are that meetings are too long, not well organized, pointless, and a waste of time. Some of the major complaints (Dressler, 1996) about meetings are as follows:

- Their purpose is unclear
- Participants are unprepared
- Key people are absent or missing
- The conversation veers off track
- Participants don't discuss issues, but instead dominate, argue, or take no part at all
- Decisions made during the meeting are not followed up

A survey of 78 project management professionals indicated that meetings were the major time consumer in their daily schedule, those meetings taking up an average of slightly less than 2 hours per day. "The survey showed that the participants attended over six meetings per week. In those meetings an average of 25 percent of the time was spent engaging in off-the-track banter, generalized complaining, lobbying for pet causes, and other unproductive activities. When respondents were asked, 'What do you think contributed to the eventual non-productiveness of the last meeting you attended?' the three most prevalent responses were (Dinsmore, 1990, 139-140)

- Improperly planned meeting (no written objective or detailed agenda)
- Inept leadership
- Undisciplined participants"

When experience with meetings is not good, people can actually get a negative conditioned response to the very word *meeting*. Yet, meetings can and do play an important part toward facilitating progress and promoting understanding among people—a key to successful management and leadership. How can this be accomplished more effectively?

Rules and Suggestions

To plan for and to conduct effective meetings, project managers and meeting participants generally need to follow ten basic rules. When used *appropriately*, these rules lead to success. They are founded upon solid principles of management and good human relations.

Rule One. *Clearly identify the meeting objective. And, if the meeting is formal, write the objective and distribute it to all concerned in advance.* This is a vital, and perhaps the most important, rule for conducting effective meetings.

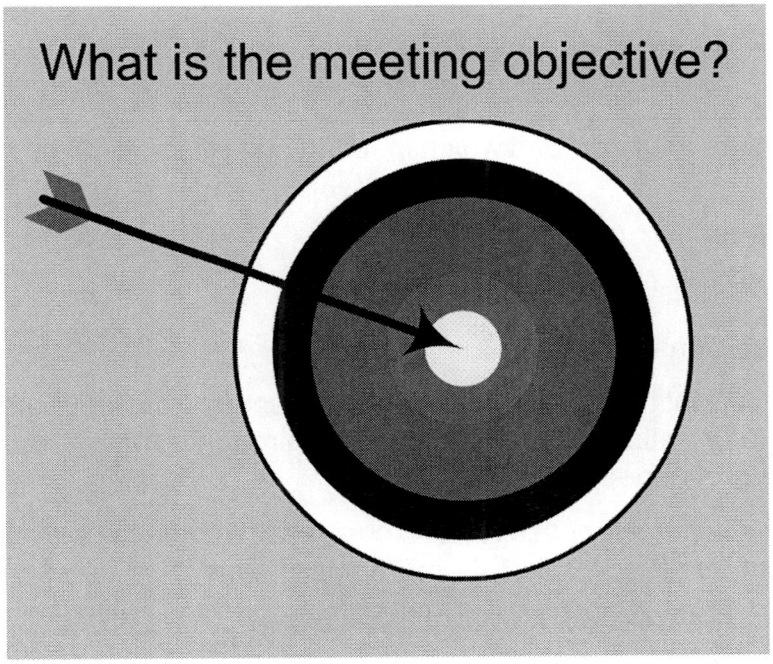

FIGURE 1. Clearly Identify the Objective of Every Meeting

Each semester in my graduate project management courses, I ask my students if they are given the objectives of the meetings they are asked to attend. For every one hundred students, I normally have less than five who are given the meeting objectives. This is disturbing when one considers that the majority of my students are full-time employees in industry, where productivity is of great concern.

When deciding to hold a meeting, a leader must determine among other things why the meeting is important and what is to be accomplished. The leader should ask, "Is this meeting necessary?" and "What is the objective or purpose of the meeting?" Answering these questions will essentially answer whether it is also necessary to meet.

Though the meeting objectives might seem obvious to the leader, they frequently are not obvious to the participants. If the meeting objective is obvious, then it should not be difficult to write that objective; yet this could indeed be a challenge to many meeting participants. If you can't write down the objective, you don't truly understand it.

> **Principle:**
>
> *If a meeting doesn't have a clearly understood purpose, don't have it! If it is going to waste people's time, either change it or cancel it!*

Some generic areas around which meeting objectives might be formulated include, for example,

- Solving problems
- Brainstorming new ideas or solutions, etc.
- Making decisions
- Sharing information (this can be a weak reason to hold a full-group meeting, since much information sharing can be done through e-mail, personal conversations, or sub-group meetings)
- Giving feedback (again, this too can be a weak reason to hold a full-group meeting, as much feedback if likely to more appropriate on a one-on-one basis)
- Planning
- Selling
- Teaching
- Conducting design reviews

Before we leave the first rule for effective meetings, there is an important point to understand. Objectives are much more effective if all key people involved agree generally on those objectives. Taking time in advance of the meeting to clarify and reach agreement on meeting objectives with all key players can go a long way toward avoiding difficulties in the meeting itself. A few advantages of doing this include

- Everyone is focused on what is to be accomplished
- Participants are less likely to come with hidden agendas

- The meeting agenda will more likely be clearly oriented to accomplishing the meeting objectives
- Agenda items not contributing to the meeting objective can and should be dropped
- The agenda items can be more easily prioritized
- The participants have a greater sense of purpose and understanding—they will behave more like a team

Rule Two. *Clearly define the meeting agenda, focused solely on accomplishing the meeting objective.* For formal meetings this agenda normally should be agreed upon in advance by the meeting participants, along with the meeting objective.

The meeting agenda should not include items that do not contribute to the meeting objective. Usually items that do not contribute to the meeting objectives can easily be communicated by some other means, such as one-on-one interaction, e-mail, or telephone conversations. Many such items also do not involve all the meeting participants anyway. Invite only those who will benefit directly from, or contribute directly to, the meeting.

Effective agendas include not only the topics to be discussed, but also who is expected to lead each topic's discussion or presentation and how much time is allotted for that agenda item. Allocate and prioritize time based on importance first, and then urgency—if an agenda item is not important, drop it. Urgency does not equal importance. As Covey (2004, 150-151) explains, "*Urgent* means it requires immediate attention…Urgent things act on us…We react to urgent matters…*Importance*…has to do with results. If something is important, it contributes to your mission, your values, [and] your high priority goals." For an excellent discussion on the difference between *urgent* versus *important*, see Covey (2004, 149-168).

Use wording to describe clearly the agenda items so that participant understanding is well focused. To further clarify an agenda topic, the leader might want to consider writing down the desired expectations next to each agenda topic. For example, if an agenda item is "Forecasts of business income," the leader might add the expectation note that "Please be prepared to decide as a result of these forecasts whether to increase production of XYZ." Making expectations clear will greatly enhance the effectiveness and efficiency of the meeting.

The items on the agenda should be prioritized with the most important items being covered as early as possible. Then, if the meeting appears to be running behind, the less important agenda items can be postponed to a later meeting without serious consequences.

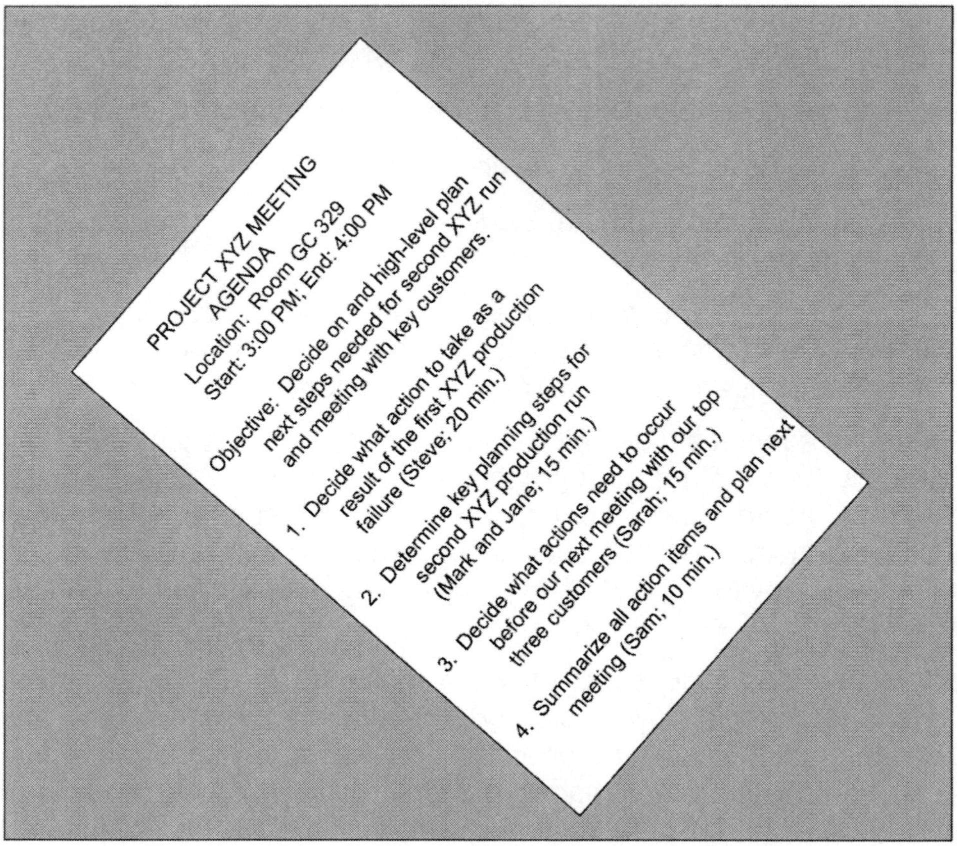

FIGURE 2. Clearly Define the Meeting Agenda to Accomplish the Objective

Be sure to schedule enough rest breaks as part of the agenda. These breaks should occur at least every one and a half hours or less, as longer periods between breaks will tend to decrease meeting effectiveness. Ten-minute breaks are usually sufficient, and should be no more than fifteen minutes except in unusual circumstances. Start promptly at the end of the break!

When distributing the agenda, include any needed background or supplementary reading material and data that will help participants prepare effectively for the meeting. Do not include material that does not directly contribute to the meeting objective.

Rule Three. *Have clear start and end times for the meeting, and treat these as a contract.* Most meetings have a clear start time. It should be firm—start on time! Doing otherwise will lead to poor meeting behavior and punish those who try to be on time.

Unfortunately, many leaders do not plan meetings to have a clear end time. This is a mistake made by meeting leaders who are not sensitive to the courtesies that ought to be shown to other participants in the meeting—their time is important and they, too, need to plan their day. To do otherwise will lead to frustration for those involved.

Once the meeting agenda has been determined and it includes time allotments for each agenda item, there should be little difficulty in deciding on when the meeting should end. Wise meeting planners include a little "contingency time" toward the end of the meeting; but this is *not* for sloppiness—it is for unforeseen events which honestly could not have been planned for. Of course other factors must also be considered such as travel schedules, meal times, and so forth. But, once the schedule is set, stick to the overall schedule. This means don't overrun the end time—you have a contract, treat it that way. On the other hand, few people, if any, will mind if the meeting ends early.

FIGURE 3. Have Clear Start and End Times, and Treat as a Contract

This does not mean that individual agenda items cannot be adjusted mid-stream. The agenda is simply a plan to reach the meeting objectives. The plan is your servant, not your master—it is there to help you! Some adaptation of the agenda might be needed to reach the objectives. Caution is in order, however—the agenda should not be so flexible that control and discipline are lost and the meeting objectives are not reached.

A final note on the time contract: It is common for the meeting to approach the end time with agenda items still to be covered. As pointed out already, the consequence of dropping off the last few agenda items will be small if the agenda has been prioritized. Still, there is frequently a temptation, especially among meeting leaders, to feel that they must get through the agenda and, therefore, they don't want to quit. The decision to run over time is *rarely* a good one. Some consequences of running overtime include

- Loosing participant attention
- Hastiness leading to poor decisions
- Loss of trust for future time contracts
- Ill will toward the leader

Sometimes we feel that most of the participants will not mind staying beyond the stop time, and we can let the one or two people who need to leave do so. While this may indeed be the case, remember that there is a reason for all participants to have been invited, including those one or two who can't stay. Otherwise they shouldn't be there. When you tell someone that it is all right for them to leave and the rest of the group will continue, you basically say, "We don't need you," or else, "We no longer need your input." Even if you state that you don't want them to get these messages, they frequently will.

This can cause feelings of loss of respect and can cause resentment, even though these results are not intended.

People who are team players will want to be involved with the discussions and decisions—don't leave them out. Conversely, the remaining team members might well feel resentment at the pressure to stay in the meeting when others are allowed to leave. It is nearly always better to keep the time contract and to schedule another follow-up meeting if it is truly needed. There may be rare exceptions, but the key word here is *rare*.

Rule Four. *Be prepared for the meeting, both in content and in process.* It is the responsibility of the meeting leader to assure that the meeting will be valuable for the participants. Proper pre-meeting planning and preparation for both content and meeting process are essential steps leading to success in making the meeting valuable.

First, *content preparation*. The agenda should be issued enough in advance so that each participant will have adequate time to be content-prepared as needed for his part. Participants should be expected to come prepared with any useful visual aids, handouts, demonstrations, reports, data analysis, and so forth, as appropriate. Use of visual aids should be encouraged, since they will help participants to better communicate and to stay focused on their assigned topic without rambling. In addition, visual aids usually greatly help in understanding.

Meeting participants should expect each other's assignments to be completed on time, and they should be prepared themselves. The expectation of disciplined preparation should be taught and used well. However, the leader should remember that this is a two-way responsibility and that he must do his part in working with the other participants in advance in a win/win approach so that these expectations can be achieved—help them prepare in any way that is needed. The leader especially must be prepared or the others will quickly see hypocrisy and often choose to follow the leader's bad example.

Additional step to take before the meeting, which will improve content preparation, include

- Consider the need for some participants to make *pre-presentations* to key decision makers before the meeting; some presentations might need to be made before the meeting to key people to prevent embarrassing surprises, to enhance time for understanding and communication before the meeting begins, and to get early feedback and buy-in. Sometimes this is called *pre-positioning*. It can be quite powerful in making meetings more effective. Meeting energy will be more focused, decision-making will be more efficient, and deadlocks will be fewer or overcome.

- As needed, determine and clarify the roles and responsibilities of all participants. This could be done one-on-one between the leader and each participant before the meeting. Some example roles and responsibilities are given in Table 4.1.

- Engender expectations among meeting participants that the meeting success will be measured mainly by how well the meeting objectives are achieved in a matter satisfying to the participants.

- Anticipate potential interpersonal communication challenges and work toward alleviating these *before* the meeting.

- Assure that all handout material is prepared in an organized way for all participants.

Second, *process preparation.* Process preparation involves assuring adequate facilities and supplies are provided to make a comfortable meeting, as well as being prepared to run the meeting in a pleasing way. There are many considerations with respect to being process-prepared for the meeting.

You should plan the meeting in a way that will maximize participation toward reaching the meeting objective; when possible, consider trying to decrease presentation time and increase time where all can participate together.

The following process items, as appropriate, should be checked well enough in advance of the meeting so that changes can still be made as needed.

- Make meeting room arrangements, including, but not limited to:
  - ✓ Reserve the right-sized meeting room and additional rooms as needed
  - ✓ Assure room temperature is properly adjusted and will not be distracting; find out *in advance* how to get it adjusted if needed
  - ✓ Assure seating arrangement fosters participation; can squeaky seats be oiled?
  - ✓ Test visual-aid equipment in advance to assure it works properly; focus projectors; check visibility from all participant locations
  - ✓ Assure miscellaneous supplies are available, such as tape, markers for white boards and flip charts, sticky notes, pencils, notepads, extension cords, flip charts, video-tape player, lap-top computer, push-pins, name-tags, microphones, projectors and screens, erasers, and pointers
  - ✓ Arrange for refreshments as appropriate, well in advance and then confirm
  - ✓ Assure adequate trash containers are available
  - ✓ Check for adequate lighting and learn to adjust it as needed
- In advance of the meeting, not during, arrange for someone to take minutes and clarify expectations for those minutes
- Are there other environmental distractions that could cause participants difficulty in paying attention or make the meeting seem too formal and not relaxed? Noise from adjoining facilities or from in-room heating and air-conditioning equipment, for example, can be unexpectedly detrimental.
- Understand the decision-making authority of the participants; find out who has authority relative the meeting objectives and find out how the group can tie into that authority quickly and effectively if need be.
- Understand what financial arrangements need to be made for the meeting and how these will be executed. Then, make sure the arrangements are made.

Table 1: Some Example Roles and Responsibilities

Example *Group-Member* Roles:

- Be prepared; fulfill all assignments on time
- Empathically listen; Communicate, communicate, communicate
- Help plan group work
- Set objectives for and plan work in alignment with the group objectives and plans
- Help establish and then obey group conduct rules
- Actively contribute toward good teamwork
- Be clear, concise, and don't dominate; but be a valuable contributor

Example Leader Roles (in addition to Group Member):

- Lead, but don't dominate
- Be responsible for meeting objectives and agenda
- Complete all pre-meeting arrangements
- Build teamwork environment; praise contributions
- Work to empower group members
- Set the example in true communication
- Lead the team in reaching the meeting objectives
- Be responsible for meeting follow-up, help people account for responsibilities

Example Minute-Recorder Roles (in addition to Group Member):

- Summarize the essence of important discussions, decisions, and reasoning
- Avoid direct quotations with names
- Clearly summarize discussion perspectives, to avoid re-hashing of these points
- Issue minutes promptly to all invitees, following approval by the leader

Example Facilitator Roles, sometimes the Leader, or another group member:

- Help plan the meeting to maximize participation and reaching meeting objectives
- Seek to decrease presentation time and increase team participation time
- Arrive early; keep the discussions on track; eliminate side conversations
- Empathically listen and help others to do the same
- Value the differences in people; then protect people from being attacked
- Help establish and enforce meeting rules agreed-upon by the group
- Facilitate processes for taking needed actions
- Assure clarity of decisions and assignments, who will do them, when they are to be completed, and to whom to report their completion
- Help leader maintain proper meeting environment and assure all participate
- Be the timekeeper
- Help assure that the minute-recorder is recording the needed information
- Make an "Issues Parking Lot" flip chart with issues to discuss later
- Lead a brief meeting-process evaluation: what went well; what to improve

- Finally, clearly understand how you will conduct the meeting and deal with possible challenges. To help prepare for this *process* part of meeting planning, consider:
  - ✓ If this will be the first time the group has met, draft proposed rules of group conduct for the group to revise and achieve consensus on during that first meeting; also, orientation kick-off materials should be prepared for each participant. If the rules have already been established at an earlier meeting, print a list of the meeting rules for all participants.
  - ✓ Decide how you will deal with unexpected issues brought up during the meeting, but which are not on the agenda—for example, will these issues be placed on an "Issues Parking Lot" flip chart to be later discussed off-line, at the end of the meeting, or at a later meeting?
  - ✓ Will you allow some potential alterations to the agenda mid-meeting, and, if so, how will this be done? For example, the group might reprioritize the agenda under your direction.
  - ✓ Consider win/win approaches to group discipline, if the need arises. If discipline is negative, you will lose the participant and you will likely lose the respect of everyone else present.
  - ✓ How will you work to reach consensus on issues during the meeting; also, what will you do if the group cannot reach consensus?
- **Rule Five.** *Near the beginning of the first meeting, the group should establish its own rules of conduct; these rules should be revisable by the group at future meetings upon request and by consensus.* Note that the word *consensus* implies general or harmonious agreement of the group members—they might not each agree one hundred percent, but they do agree willingly to go along with the group's rules toward achieving the group objectives. Note also that some of the group's meeting rules might need to have clearly defined consequences if they are broken. Some possible examples of group meeting rules of conduct are shown in Table 4.2.
- Although it is possible for groups to function without agreed-upon rules for meeting conduct, "Common, or at least harmonious, goals or purposes are...not criteria of groups, but of *effective* groups." (Hersey et al., 2001, 318)
- While considering meeting rules, don't forget that one of the primary objectives a leader must have is to build the group and to positively develop the individual group members. All meetings, and especially the first one, should strongly engender member participation and teamwork. Group and individual morale is critical to achieving group objectives. Overly oppressive rules can backfire. The group members need to feel welcome, needed, and honored to be present.

Rule Six. *Understand the fundamental meaning of the word* communication *and apply it well and continuously.* The leader and meeting members should use this understanding before, during, and after the meeting. The word *communication* means literally "the process of becoming like one or as one." By this definition communication cannot be one-way. The process of becoming like one involves much give-and-take, much effort to try to put oneself in the other person's shoes, or as Covey (2004, 237) says, "Seek first to understand, then to be understood." It is truly a coming together in understanding and purpose.

Before, during, and after group meetings this concept of communication, if applied diligently, will reap tremendous rewards toward having meetings be truly effective. This application of communication will aid a group to build true team spirit and unity in working toward and achieving a common goal. It is also a builder of human dignity and effective human interpersonal relations.

TABLE 2: Example Meeting Rules of Conduct (select less than 7)

- We arrive early; start, and end on time
- We will work to achieve the meeting objectives
- We will come prepared and stay on track
- A person missing more than two of four consecutive meetings may be dropped from the group, subject to a majority vote
- To meet, 75% of the group must be present
- We strive to accommodate all members when scheduling group meetings
- We will follow good behavior etiquette and courtesy toward each other and strive to achieve win/win results as much as possible
- Meeting minutes issue within three working days
- We practice Empathic Listening; Seek first to understand, then to be understood
- We will not dominate discussions, but we will actively participate
- We will not do or say things that discourage other members from participating
- We build on each other's ideas
- One person talks at a time; no side conversations
- No gossip
- We don't make remarks that are sexist, racist, prejudiced, sarcastic, profane, or negative to any religious, social, or other organization or person
- We evaluate things, issues, and processes, not people, unless the meeting purpose is specifically to evaluate personnel for feedback and building
- Ideas belong to the group, not the individual
- The group speaks with one voice after decisions are made; leave united
- Turn off cell phones, beepers, and alarms
- Keep your cool. If you lose it, you are wrong

Rule Seven. *Show by example and teach group members how to empathically listen to each other.* Again, as in Rule 6 above, this applies before, during, and after the meeting. Empathic listening is essential to communication as described above. It must be based on sincerely caring about the other people involved; otherwise it will be perceived to be fake or manipulative. One must seek to really care about other people. "The…listener's full attention is on the content of the message *and the intention* of the speaker." (Hersey et al., 2001, 305; italics added) Listening for the speaker's feelings is just as important as listening for the facts the speaker presents. See Figure 4.

Following are some suggestions you and your group members can use to improve empathic listening.

- *Listen for the concepts and details of the message*. Make an effort to really listen precisely what the speaker is saying.
- *Listen for the feelings of the speaker*. Try to perceive the speaker's feelings, desires, and emotions an about what is being said through the way that the message is delivered—try to really put yourself in the other person's shoes and try to see things they way they see them.
- *Respond to the feelings of the speaker*. Sincerely reflect to the speaker what you recognize and understand the feelings expressed to be.
- *Watch for the speaker's verbal and nonverbal messages*. Attempt to identify the real messages the speaker may be expressing.
- *Reflect back to the speaker the concepts and feelings you are hearing*. In your own words try to share with the speaker what you think the speaker said and felt. Listen to how the speaker responds to your reflection and interact iteratively until you both understand each others messages and feelings. (Based in part on Hersey et al., 2001, 306, quoting Rogers, 1951, and on Covey, 1989)

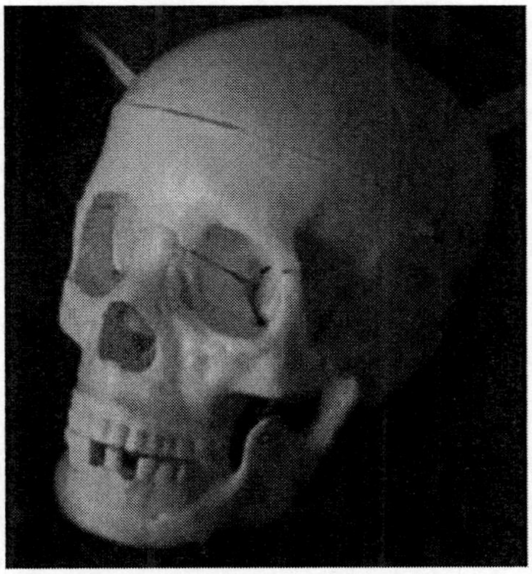

FIGURE 4. I've Been Waiting a Long Time for You to Listen to Me

The advantages of empathic listening are many. Here are a few.

- You will better understand the perspective of the other person, which will place you in a better position to cooperate,
- People will know you care about them,
- You obtain feedback,
- You gain better perspective and understanding,

- Mutual acceptance and respect are engendered,
- You will more easily achieve win/win solutions,
- You will more easily achieve synergy among the group members (Covey, 1989),
- Your group will be less likely to make costly mistakes due to misunderstandings,
- Your group will be more empowered due to better understanding and knowledge, and
- Your group will have better morale.

Many years ago I went to my manager's supervisor with a concern about how we could increase the availability of current technical literature for the engineers in our company. As I entered the supervisor's office, he shut the door behind me, sat down at a desk with me, and made me feel at ease. We had a little meeting. As I explained my concern, the supervisor took out a notebook and carefully noted my comments. He asked clarifying questions to make sure he understood. He asked for my suggestions and feelings. He summarized what he thought he heard me say and made certain that he understood correctly. Finally, he seemed genuinely appreciative that I would give him these comments and suggestions. As I left this man's office, I felt both that he had fully understood and that he cared. I also remember thinking that it really didn't matter to me anymore whether the supervisor followed my recommendations or not. Because I felt so much that he really understood me, I felt I could trust his judgment to do what was best. He had won the battle by empathic listening, whether he took any action or not.

Rule Eight. *Create a supportive and task-oriented environment during the meeting.* This rule will be accomplished as the Rules 1-7 above are followed, coupled with a dedicated effort by all meeting members to accomplish the meeting objectives. The leader and, if present, the facilitator play the crucial roles in accomplishing the benefits of this rule. They need to stay focused on the objective and care about the people! A leader who does not follow this rule will be the major cause for a miserable meeting; conversely, a leader who applies it properly will almost always have an effective meeting and one people can positively remember.

A task-oriented environment will help the meeting stay on track toward the meeting objective. Sometimes it is useful to have the team participate directly in the effort to stay on track. At the beginning of one meeting I attended, the leader handed to each member a sheet of paper with the words "We Are Off Track" in large bold letters. Anytime any group member felt the discussion was going off track, that member could hold up his sign. The result was both fun and effective.

In a supportive and task-oriented environment, meeting participants help each other, rather than hurt each other. For example, it would be inappropriate speak degradingly to or to yell at someone in front of other people. While there could be very *rare* exceptions, it is almost always not good practice to discipline someone in front of other people.

It is not appropriate to "launch missiles and throw grenades" during a meeting. Launching a verbal attack on someone in front of other people alienates them and makes a fool of you. If you do this you will lose the respect of the other meeting participants as well. If you lose your cool, you are wrong. People have been asked to participate in the meeting for a reason—or at least they have if you are following good meeting rules. Do not alienate them or they will not be able to fulfill their purpose in the meeting—they might be there in body, but they will not be there in spirit and will likely stop contributing. Thus, everyone loses, especially you!

In a supportive and task-oriented environment, meeting participants don't hold "Two meetings at the same time"—meaning that they don't carry on side conversations during a meeting. See "Don't hold two meetings at the same time." (STORY)

FIGURE 5. Create a Supportive and Task Oriented Environment

Don't come with hidden agendas to a meeting. You are out of place to do this. If you do not agree with the agenda, try to resolve this is one-on-one with the meeting leader, *well before* the meeting. Occasionally, if several of the potential meeting participants do not agree with the agenda, they could all meet in advance with the meeting leader to try to work out a compromise. In any case, when the meeting begins the time has passed to come with your own agenda.

In a supportive and task-oriented environment meeting participants sincerely phrase critiques by beginning with words such as "I wish I understood better…" or "Tell me more about how…" By doing this you might find that one of the following happens: (1) As the person explains more about their idea, they may actually convince themselves that it isn't what they initially thought, (2) As the person explains more, you might realize that it isn't what you initially thought, (3) As the person explains more, he feels so glad you are actually listening that he doesn't feel a need to pursue that idea forcefully—if you don't listen he is likely to pursue it forcefully, (4) you might find that you actually agree, or (5) you might get an new idea inspired by the other person's explanation, which idea you might not have thought of otherwise. The bottom line is to be supportive in your interactions with other people and their ideas.

Finally, in a supportive and task-oriented environment, all meeting participants should work to achieve the meeting objective(s) and not waste people's valuable time.

Rule Nine. *Close the meeting effectively.* The following hints can be very effective in closing the meeting.

- Assure that assignments and other action items are clearly defined. Use action verbs in the description of all assignments—action verbs tell what action to take and they are powerful
- Clearly decide who will do each assignment
- Clearly decide when each assignment is to be completed. If at all possible, the person who is to do an assignment should be the one to decide when that assignment will be completed
- Decide to whom the person doing an assignment will account for its status and completion—this is almost always forgotten, but can be a very helpful clarification leading to high effectiveness
- If this group is to meet in the future, have the group do the following three actions
    - ✓ Draft an objective and agenda for the next one, two, or possibly even three meetings—these are only drafts but will keep the group focused on the long-term vision of where it is going and give it a fuller sense of purpose and the need for timely performance. Note: This will be too time-consuming if built from scratch in the meeting. In advance, the meeting leader should have prepared rough drafts of these future meeting objectives and agendas for the team to revise as appropriate; otherwise this step will become disliked by the team.
    - ✓ Set the time and place of the next meeting. Do this after the objective and agenda are drafted for the next meeting(s) so that the group can identify how soon they should meet.
    - ✓ Critique the current meeting. What went well? What could be done to make it better? —This critique should not last longer than five to ten minutes or the group will not want to do it again. Keep it positive! I usually ask the group to complete the following two phrases: "I liked…" and "I wish…"

Rule Ten. *Follow up.* The meeting leader needs to build trust, accountability, and respect by following-up on assignments and action items in a caring but timely way as appropriate for each meeting member and each task to be completed. There are two objectives here:

(1) build teamwork toward accomplishing the assigned tasks and (2) build the meeting members so that they progress based on their individual situations.

Recommendations

How does one go about implementing the above rules? At first, it could seem overwhelming. It takes time and experience to implement these rules correctly. However, if meeting participants will keep the end in mind—that they want to reach their objectives in a way in which everyone wins—then they will have made the first and most important step. For many people good meeting management will be an easy and exciting step forward; for others it will be a culture shock. In all cases it is needed.

The above rules, lists, and suggestions are really tools and principles that need to be applied according to the needs of the situation and the people. Small steps at first are appropriate so that people can learn them and buy-in to the better ways. It will not take long for meeting participants to see the benefits and the need to change. To run too fast into full implementation of these rules could cause rejection due to misunderstanding and the natural tendency by many of us to hold to the *status quo*. Care deeply about your people, but persistently move forward toward helping them to have more effective meetings. Most people will welcome the needed changes.

Conclusion

Conducting effective meetings takes a concentrated effort, but the return on the effort invested will pay big dividends both during the meeting and for the long term. Objectives will be clearer and will be achieved. Morale will be higher, as will productivity. There is really little or nothing to lose by doing it right, but a great deal to gain. Meeting members will want to come back again, rather than dreading the words, "Another meeting."

References

Covey, Stephen R., *The 7 Habits of Highly Effective People—Restoring the Character Ethic,* Simon & Schuster Inc. (1989).

Dinsmore, Paul C., *Human Factors in Project Management,* AMACOM (1990).

Dressler, C. 1996. Cited by J. P. Lewis, *Team-Based Project Management,* AMACOM (1998). Cited also by Karl A. Smith, *Project Management and Teamwork,* McGraw-Hill (2000).

Gido, Jack, and James P. Clements, *Successful Project Management,* South-Western College Publishing (1999).

Hersey, Paul, Kenneth H. Blanchard, and Dewey E. Johnson, *Management of Organizational Behavior—Leading Human Resources,* 8th ed., Prentice Hall (2001).

Meredith, Jack R., and Samuel J. Mantel, Jr., *Project Management—A Managerial Approach,* John Wiley & Sons, Inc. (2000).

Rogers, Carl, *Client-Centered Therapy: Its Current Practice, Implications, and Theory,* Houghton Mifflin (1951). Cited by Hersey, Paul, Kenneth H. Blanchard, and Dewey E. Johnson, *Management of Organizational Behavior—Leading Human Resources,* 8th ed., Prentice Hall (2001).

The Importance of Excellent Definition of Project Objectives

Paper # 474824
Dr. Eldon R. Larsen
Marshall University

Effective project objectives are vital for successful project management. They tell us what we want the end service, result, or product of a project to be. Surprisingly, most project management texts go little beyond acknowledging the importance of good objectives. The *"How to"* is either missing or given only very brief discussion. This paper presents some basic principles and tools helpful in setting effective objectives. The tools help address questions around project scope, decision-making trade-offs, and tolerance between what is ideal and what is failure. These same tools aid in communication between the project team and other key players so that they all understand the intended project outcome.

Introduction

Clear, specific, effective objectives give us the best picture of a project's desired destination. "Objectives are the aims or end points toward which activity is directed." (Babcock, 1991) Someone once said, "I always wanted to be successful, but I guess I should have been more specific. I should have defined success."

When working on projects, what do the project objectives really mean? Do we really understand them? Do other people understand them the way we do? How will we know when the objectives have been achieved? Is there any room for variation if we have trouble reaching the ideal? Have key people agreed upon the objectives? What constraints are we facing? Which constraints are most important and which least important?

These and other questions are the topic of this paper as we investigate one of the most important parts of starting projects right—how to develop effective project objectives.

Tool 1: Make Objectives SMART

One very helpful tool for developing effective project objectives is the acronym SMART, described in **Figure 1**. Project objectives need to be specific, measurable, agreed-upon, realistic, and time-oriented. A weakness in any one of these areas can be either like a hidden time bomb ticking away and ready to blow up at any time, or like a cancerous problem that slowly grows until it either kills the project or requires it to have major surgery and expensive life-saving resuscitation. Of the many variations on what the letters in the acronym SMART stand for, the author believes the one presented in **Figure 1** to be most useful. It is based on Smith (1991), but the author has expanded the definitions for clarification.

S	**S**pecific:	Targeted, focused, clear, concrete, understandable, documented, not abstract, not general, not overly complex
M	**M**easurable:	You can tell when you've reached it, targets are clear, tangible, verifiable, and quantified if at all possible
A	**A**greed-Upon:	Everyone involved is aligned; owner, senior management, project leader, project team, and other key people have approved and committed, preferably in writing
R	**R**ealistic:	Challenging, but attainable; not impossible to achieve nor viewed that way; consistent with resource and time constraints; not too rigid to accommodate changing priorities
T	**T**ime-Oriented:	Clear time frame for project completion

Figure 1. Objectives Should Be SMART

Specific. A project objective that is *specific* will be targeted, focused, clear, concrete, understandable, documented, not abstract, not general, and not overly complex. What we are trying to do with the objective statement is to be very good artists and to paint a very clear picture for everyone involved with the project. That artistic picture should not be like the impressionistic artistic style such that the words appear pleasing from a distance, but when you get close up there is not enough detail to understand what is intended. Instead the style needs to be more along the realistic artistic style, where both far-away and close-up examinations provide good understanding.

Part of being specific is to put the objective in writing. Sometimes for small projects we feel this is not worth the trouble. However, the act of putting an objective in writing forces us to clarify it. If you can't write it down, you don't understand it.

Measurable. The project objectives also need to be *measurable*. How will we know when we have achieved the desired outcome? We need to be able to measure this in some specific way. The objectives need to be clear, tangible, verifiable, and quantified if at all possible. "You can't prove a project's success if there aren't any measures for it." (Frye and Mason, 2001)

I once worked on a research project for approximately four years with a very fine team of people. We met regularly and were very familiar with what was happening on the project. Yet, after four years I started hearing comments in the team meetings that in effect said, "Are we finished now?" Some members thought the answer was "Yes." Others thought "No." Upon examination of the objectives, we could not tell. The objectives were neither clear enough nor measurable enough to allow us to make that decision—after four years!

In making objectives measurable, one key is to decide both *what* you want to measure and *how* you will measure it. The *what* and *how* both need to be specific enough so that everyone knows when they are met. Sometimes the measurable characteristics of the project objectives are called metrics. Regardless of the terminology, we need to answer the *what* and *how* measurement questions.

For example, one of the objectives of a project might be to build a sailboat that is thirty-six feet long. One part of the *what* might be the dimension thirty-six feet. *How* we can measure the objective might be with a tape measure. This example is fairly easy.

Another example might be that we have an objective for the people at our family reunion to enjoy the food provided. The *what* is a little harder to quantify here, to enjoy the food. The more difficult question is *how* to measure that enjoyment. We could have an end-of-reunion critique meeting in which the food is discussed; we could give everyone a written survey; or, we might come up with another more creative method of measurement. The point is that if we do not define how we will measure the objective, there is a good chance it will not be measured and we will not be able to determine if the objective has been achieved.

Let's look at one last example of a measurable objective. A hotel manager might have as an objective, "To increase the hotel owner's satisfaction with my management performance." Although this is a very nice desire, the objective is neither very specific nor very measurable. First, the hotel manager needs to define better what would affect the owner's satisfaction. Second, the manager needs to define better what the word *increase* means, as this could be anything from a very small change to a very large one. What does the owner want? What can the manager provide? As an example, perhaps a more specific and measurable objective might be, "To increase the average hotel room occupancy from the past value of eighty-five percent to a new value of at least ninety-five percent during the months of June through August of this year." This revised objective is much more measurable and specific.

Three cautions are important when making objectives measurable. First, watch out for common words that sound desirable, but are not specific and not measurable. A few examples are the words *good, excellent, pretty, high quality, long lasting, preferred,* and *successful*. To correct the problem with these words, simply express in quantifiable terms what you really mean. For example, the phrase "We will provide excellent help-desk service," might be changed to something like, "We will answer the telephone for ninety-five percent of all help-desk calls within 20 seconds and we will provide solutions acceptable, according to the customer, to at least ninety percent of the problems within one working day."

The second caution is to be careful not to skew your objectives just so you can easily measure them. This can lead to disaster just as badly as can the original immeasurable objectives. The challenge is to re-express your real objectives in a way that they are both clear and measurable as much as is realistically possible.

The third caution is to be very careful that what you choose to measure, and how you measure it, do not cause unwanted performance side effects. Sometimes while driving desired behaviors, we inadvertently drive new, unwanted behaviors. There is an old saying that you get what you measure.

Agreed-upon. The project objectives need to be *agreed-upon*. This means that everyone involved with the project is aligned and that, preferably in writing, the project owner, senior management, project leader, and other key people have approved and committed themselves to the project objectives.

Some project owners give the project team a set of objectives and simply expect them to go to work and achieve them. While this can work in some cases, experience shows that participant buy-in is usually very important for effective organizations. This is true as well for project organizations and teams.

By allowing a project manager, and where practical the project team, to have a chance to work with the project owner on the definition of the objectives, a far better set of project objectives is probable, and participant buy-in is more likely and much higher. In the long run, this extra effort pays big dividends and is more likely to produce win/win situations for the owner and the project participants.

When an owner gives a team project objectives with which they might not fully agree, and without any way to iteratively improve those objectives through two-way communication, that owner unintentionally is also often asking for ineffective results, inefficiencies due to poor morale or poor understanding, and possibly even project failure.

If used properly, the tools and concepts presented in this paper can actually be of great help in improving communication between the key people involved with a project and go a long way to facilitating agreement upon objectives.

Realistic. Project objectives should be *realistic*, meaning that they are challenging but attainable. They should not be impossible to achieve nor be perceived that way. They should be consistent with resource and time constraints. And, order to accommodate changing priorities when essential, they should not be too rigid. "Generally, the more difficult the goal, the higher the level of performance. But…if and when a goal is perceived as so difficult that it's virtually impossible to attain, the result is often frustration rather than achievement." (Gibson, et al., 1997)

One caution is worth mentioning with respect to making objectives challenging: Occasionally a project owner will allow conflict to exist within the project objectives, perhaps even intentionally, thinking that higher achievement will result. This is normally not a positive motivation for teams. Gibson, et al. (1997) report that a recent study "examined conflict as a goal-setting variable. Both a laboratory experiment and a field study of college professors suggested that conflicting goals can lead to decreases in performance…This research points out the need to pay attention to the total situation."

In summary, project team members will be better motivated both as individuals and as team members when objectives challenge them in a positive way, but do not overwhelm them due to either real or perceived lack of attainability.

Time-oriented. Finally, project objectives need to be *time-oriented*, meaning that there is a clear time frame for project completion. Sometimes, particularly in research and development, teams are given project objectives without any idea as to what the time-limit expectations are. Although there are situations where time limits are difficult to define, even in those cases an approximate time *range* in which a project needs to be completed should be describable. For example, a research project outcome and endpoint might be uncertain, but funding cannot continue indefinitely. In such a case, we might have to say something like "This project is to be completed within the next two to three years." Although some people might feel uncomfortable with this flexible time range, it is much better than saying nothing! And when there is high uncertainty, an ending time *range* in not inappropriate. That range can be narrowed down as time progresses and the project develops. We will talk more about *time* as part of our discussion of the triple constraints, below.

Tool 2: Make Full Use of the Triple-Constraints

Another important concept and tool is that of the triple constraints, illustrated in **Figure 2**. A constraint is something that imposes a limit or boundary. In project management, we normally speak of three major constraint criteria: the triple constraints. Although there are some variations of this concept, the author has found the one shown to be generally very useful, the three general constraints on the project being *time*, *resources*, and *performance* criteria. Let's examine each one.

Time Constraint. The *time* constraint, already introduced briefly above, is relatively easy to understand; at least that could be our impression. However, perhaps it is this impression of easiness that causes people occasionally to overlook adequately describing it when they write project objectives. The time constraint is very important. It basically involves identifying when the project is to be completed, at least roughly. Associated with the time constraint is how long it will take to do the project—the project duration. It is not the purpose of this paper to go into the many tools useful for planning a project and estimating project durations. The point here is that it must be done.

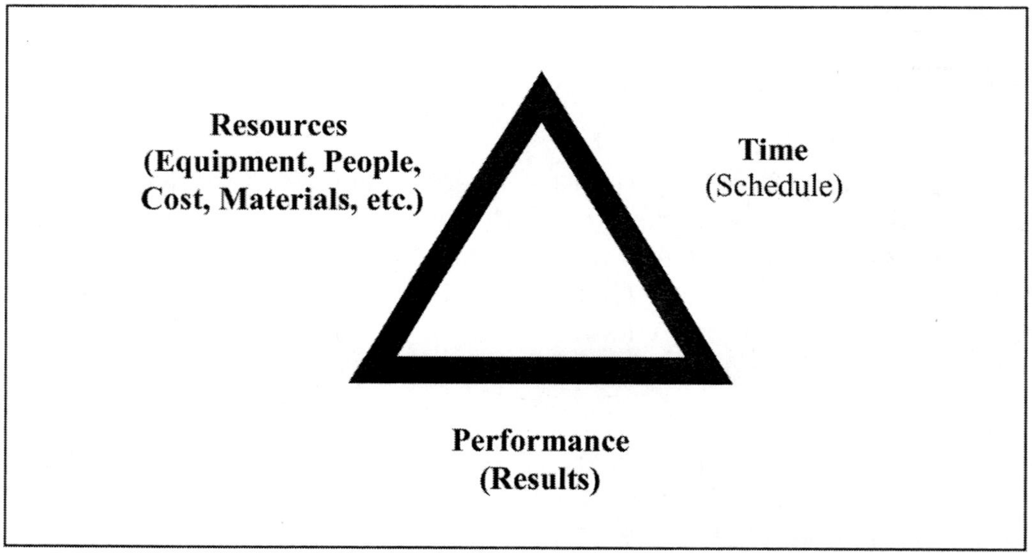

Figure 2. The Triple Constraints

It is not uncommon for the owner of a project to give a deadline for a project without due consideration as to whether the deadline is reasonable. Normally, to determine an initial, realistic time limit estimate, at least some high level planning of the entire project is needed. This initial time estimate can be refined later, and should be, but setting time limits without at least high-level overall project planning is not only unwise and likely to cause much frustration, it can also be disastrous to the project.

There are cases where a deadline is fixed due to reasons beyond the control of either the owner or the project team. For example, legal or regulatory implications, market competitions, or business and financial constraints might dictate a deadline. When this is the case, it is still very important to do at least a high-level plan to see if the deadline has a realistic chance of being achieved. Not doing this can cause a large amount of wasted money, effort, and time.

Resources Constraint. The *resources* constraint is a little more complex. Sometimes it is simply stated as cost or budget, but in reality it includes also items like people, equipment, materials, facilities, etc. When a project owner and/or project manager focuses only on budget as the resource constraint, the other resources of people, equipment, materials, etc., can be neglected until problems have had a chance to develop to the point that project execution is hampered. The message here is that all the project resources and their interactions need to be considered as a constraint on the project.

Performance Constraint. The third general constraint is *performance*, or results. It includes many possible specifications and is sometimes, but not consistently, referred to as project *scope*. Sometimes performance is defined as the project technical objectives, but in reality it can include much more, or, depending on the project, it could have nothing to do with anything technical at all. Examples of some of the items that might be considered for inclusion in the performance are given in **Figure 3**. Care must be taken not to neglect these possible aspects of performance when writing the project objectives. Some of the items in **Figure 3** were taken from Bitz (1991).

- Deliverables
- Quality
- Production rate
- Product specifications
- Customer satisfaction
- Stakeholder satisfaction
- Labor relations
- Safety
- Health
- Environment
- Work processes
- Quantity
- Reliability
- Operability
- Modularity
- Flexibility
- Maintainability
- Community relations
- Corporate image
- Product cost

Figure 3. Some Example *Performance* Constraints

Interrelation of the Triple Constraints. Note that the triple constraints are not independent of each other. A needed change in the time constraint can affect resources. A change in the resources can affect our ability to meet a time constraint, or cause a need to change that time constraint. Changes in performance criteria due to new knowledge or new situations can cause a need to adjust both the resource constraint and the time constraint. Therefore, use caution when modifying constraints to assure that the effects on the other constraints are well understood and accounted for.

Triple Constraints Are Project Drivers. There is a general understanding of the triple constraints by many project managers, but one of the most important uses of the triple constraints is often either unrealized or neglected—the concept of the triple constraints being the project drivers. By *drivers* we mean: Which one of the triple constraints (time, resources, or performance) is most important, i.e., the *primary driver;* which one is least important, i.e., the *weak driver*; and which one is in the middle, i.e., the *middle driver*? (Dobson, 1993) Do we really need to care about this? Aren't time, resources, and performance all important? Do we really have any choice? The answer to all of these questions is, "Yes."

The truly vital question to understand when discussing the importance of the drivers is, "Why do we care?" We care because at some point in the project we are going to have to make vitally important decisions with respect to the project, and the correct decisions are going to depend on the driver priority or importance relationship.

All three of the triple constraints are important for all projects, but they are *not* all three *equally* important in any project! The trick is to determine which constraint is most important, which least important, and which is in the middle for the particular project concerned. How do we do this?

Well, we could ask the project owner which driver is most important, time, resources, or performance. In response the owner is likely to say something to the effect, "What do you mean by asking me this? They are all essential and I expect you to give me them all. Now go back to work." This approach will usually fail. The problem is that even the project owner usually does not think in these terms. You will have to help him.

One way to do this is by asking the project owner a variety of trade-off questions. It is best first to patiently explain to the owner what you mean by the triple constraints. Tell the owner that you know that each of the constraints is very important. Then, tell the owner that because you really want to manage the project so that he or she is satisfied, there will come times during the project that you will have to make important decisions, and you want to do your best to make the right choices. Then, give the owner some example general trade-off questions that you might be faced with. Don't get too detailed; stay general—you probably don't really know at this point what problems you will run into and what decisions you might face. Right now we are talking hypothetical. What you are trying to do is to weigh each possible pair in the triple constraints—time vs. resources, time vs. performance, and resources vs. performance.

For example, you might say something to the effect, "Although we don't want this to happen, suppose as the project progresses we realize that in order to finish by the project due date, we must hire some people to work overtime, thus increasing the cost a little, and causing us to go over budget a little. Would you rather have us do that or would you rather we stay on budget and finish the project late by a week or two?" The project owner normally will be able to give you an answer to this question that gives you a feeling for whether resources or time is more important—he might even explain the reason, which would be very helpful insight that you did not have earlier.

Then, you could ask a trade-off question between the performance and the time; then, the performance and the resources, and so forth. You should ask enough different questions so that you get a clear feeling for the proper order of the triple constraints: which is the primary driver, the middle driver, and the weak driver.

Finally, state clearly to the owner what you have perceived to be the constraint priorities and get verification that the owner is in agreement. Putting it in writing would be best.

Driver Cautions. There are several cautions to be aware of when determining the triple constraint drivers.

First, there are things that can develop during the course of a project that cause the order of the drivers to change. Watch out for this possibility. Frequent communication with the owner and the other key project personnel is vital to help you to constantly be aware of which constraint is most important, etc. A change in driver priorities should only occur when absolutely necessary, and then only with agreement among the key decision makers on the project, such as the owner, the project manager, and other key leaders. Such a change should always be rapidly communicated to all people involved with the project.

Second, you should not purposely plan the project with the intent that the order of the drivers will change. For example, at the beginning of the project you don't say that for the first half of the project we will have time be the primary driver and then for the second half we will have performance be the primary driver. If you and the owner have agreed that performance will be the primary driver at the end of the project, then you plan the entire project from the beginning as if performance is always the primary driver. Then, if the primary driver must change, it changes, and you re-plan accordingly.

Third, you might find that two of the three drivers are both extremely important, with the third being less important. Does this mean that you should declare a tie and have two primary drivers and one weak driver, with no middle driver? The answer is "No!" Ask more trade-off questions between the two drivers that you think are tied. Even though one of them is to be designated a middle driver, this does not mean that it cannot be a very, very close second to the primary driver. It can. But when problems arise in the project and decisions have to be made, you must know beforehand which driver is on top.

Finally, as already discussed above, the triple constraints, and thus the drivers, are not independent of each other. Always be aware of how a change in one affects the others.

Tool 3: Project Tolerance Diagrams

Let's suppose the owner of your project specifies that the project is to be completed by July 4, 2003. Our impulse might be to treat this as the deadline and do everything in our ability to reach it, even if it means hiring people on overtime or cutting a few corners in performance to do it. On the other hand, we might tell ourselves that the owner just pulled July 4 "Out of the air," so to speak, as the date that he would like to have, but it is not essential and that sometime soon after July 4 should be all right—it is ideal but not really that essential. Which assumption is correct? Well, it could be either one. Do we really need to know this? Absolutely! The answer could mean the difference between project failure or not.

A tool to help us with this problem is called the *Tolerance Diagram*, shown in **Figure 4**.

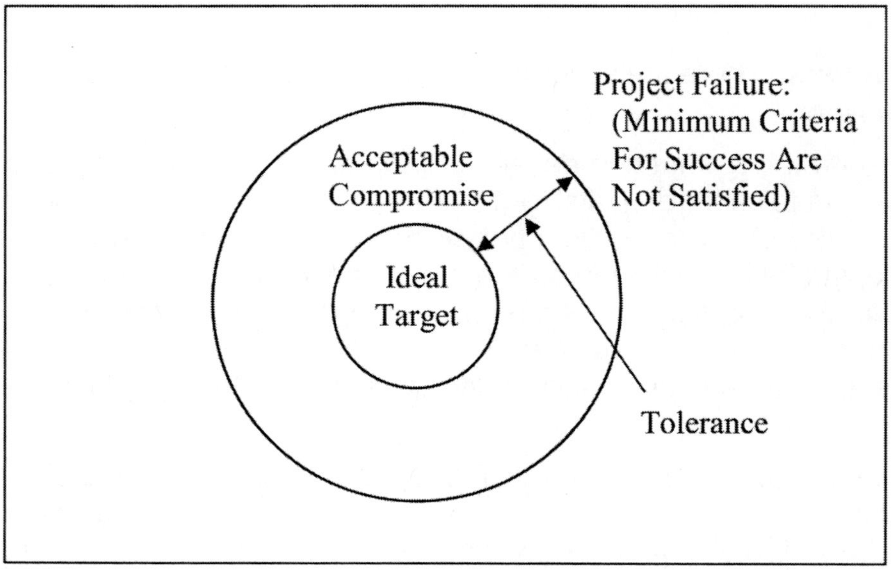

Figure 4. A Simple Tolerance Diagram

Three zones exist on the Tolerance Diagram in **Figure 4**: The ideal target zone, the acceptable compromise zone, and the failure zone. The ideal target zone describes what we would really like to have, if at all possible within the constraints on the project. The acceptable compromise zone is what we must have, even if we can't have the ideal. If we can't have at least the acceptable compromise, then we move into the failure zone and the project must terminate, or another alternative must be determined, including the possibility of a different or a significantly re-defined project. We call **Figure 4** a Tolerance Diagram because it illustrates how much tolerance or flexibility there is between our ideal desires and what might be considered failure. The primary driver will tend to have the least flexibility or tolerance, while the weak driver will tend to have the greatest flexibility or tolerance.

Let's look at an example. Suppose our only family car has broken down and this time it really is just not worth it to get the car fixed—we need to purchase a replacement. We have certain limitations or constraints, one of which is the money to purchase a car. It would of course be nice to have a new red Ferrari, but the cost is out of the question—it would cause failure in our credit rating within a couple of months. What we need is car that will reliably start and will get us to work each day—this we must have. After some financial figuring and some emotional stretching, we decide that we would like to spend no more than $5,000 on the car—this is our ideal target, a cost less than $5,000. We also decide that if we really can't find a car for a cost less than $5,000 that meets some minimum reliability standard (reliability is a performance constraint), we would be willing to go as high as $7,000, but absolutely no higher. Our acceptable compromise for the money constraint is therefore a cost that is between $5,000 and $7,000. Failure is to have a cost greater than $7,000.

People familiar with various problem-solving techniques will note some similarity between the one-sided Tolerance Diagram described above and the suggestion by Kepner and Tregoe (1981) that there are *wants* and *musts* that need to be identified when trying to evaluate alternative solutions to problems. Although Kepner and Tregoe's wants and musts are not exactly the same respectively as the ideal target and the acceptable compromise zones of the Tolerance Diagram, the detailed analytical approach they use to weigh various alternatives could conceptually be applied to the Tolerance Diagram, and interested readers are referred to their work.

Figure 4 is a *one-sided* Tolerance Diagram, meaning that going from best to worst is one-directional. Rather than the simple one-sided Tolerance Diagram, some project specifications will require a *two-sided* Tolerance Diagram, as illustrated in **Figure 5**. In this case the ideal target is a range in the middle of a project specification or constraint. On either side of this ideal target range are ranges deviating to one side or the other, which are still acceptable, but not ideal, giving therefore two acceptable compromises. Going outside of these two acceptable compromise ranges causes the project to move into unacceptable ranges, which again we call failure.

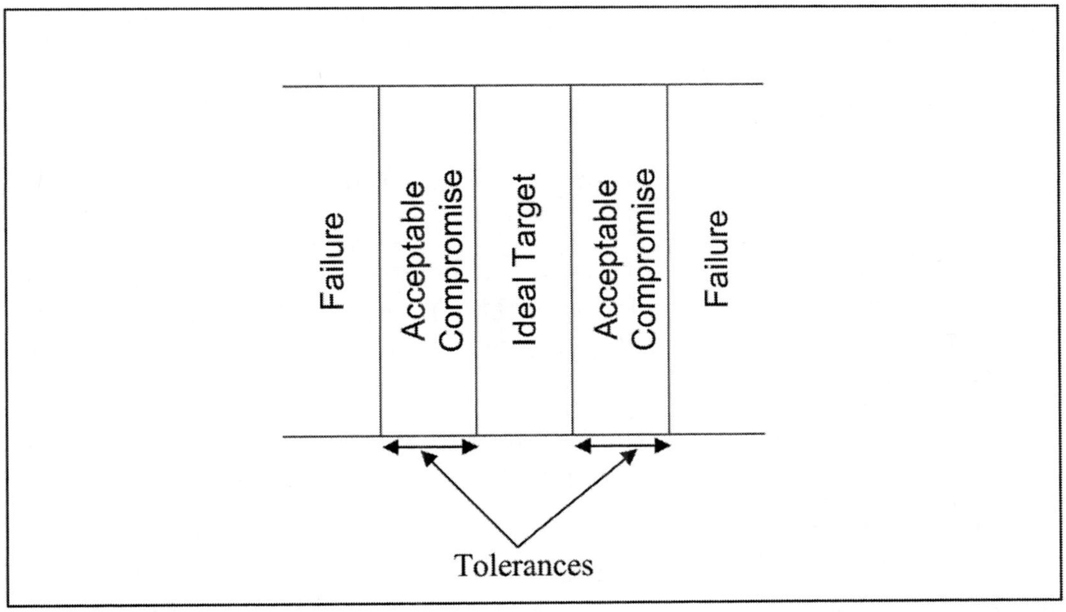

Figure 5. Two-Sided Tolerance Diagram

Let's look at an example of a two-sided Tolerance Diagram. Suppose that we are trying to make a new polyethylene polymer product that meets a certain density specification. We might want to have the polymer density be 0.950 g/cm^3. It will not be possible consistently to make that polymer with a density of exactly 0.950 g/cm^3, so instead we might specify that our ideal target range is 0.948 g/cm^3 to 0.952 g/cm^3. Although this is the ideal target, if we really must we could accept as a compromise polymer as low in density as 0.946 g/cm^3 or as high as 0.954 g/cm^3. Therefore, one acceptable compromise range would be 0.946 g/cm^3 to 0.948 g/cm^3 and the other would be 0.952 g/cm^3 to 0.954 g/cm^3. Any density less than 0.946 g/cm^3 or greater than 0.954 g/cm^3 would be failure.

Care should be taken when developing project objective specifications and identifying constraints to understand whether we need a one-sided or a two-sided Tolerance Diagram. Many projects will use both, some specifications or constraints using one-sided Tolerance Diagrams and the other specifications or constraints using two-sided Tolerance Diagrams.

Generally there should be a minimum of three Tolerance Diagrams for every project: one for each of the triple constraints of time, resources, and performance. However, usually you will need more than three. This is because the resources constraint and the performance constraint each have several aspects to them.

Looking at performance for example, we might have one Tolerance Diagram for some quality specification, another Tolerance Diagram for meeting some safety objectives, and yet another Tolerance Diagram for a new product production rate.

The author has seen an approach similar to Tolerance Diagrams receive very favorable reviews by the project team on a very large successful industrial capital product. The team reported in an end-of-project review that it had never before seen better-defined project objectives.

One of the arguments sometimes given against using the tolerance diagram concept is that perhaps people will not work as hard if you tell them the acceptable compromise in addition to the ideal target. This argument is weak for the following reasons.

First, if people have worked for you before, they have already figured out if you have habitually told them solely the ideal target, almost always to compromise it later on. In this case it is better to tell them the acceptable compromise up front rather than for them to make one up in their minds and be continually second guessing you.

Second, if you have a proper reward structure that rewards project teams for attaining the ideal objective as often as possible, they will respond to the rewards.

Third, people learn to trust you better when they know you are fully communicating with them and telling them the whole story—telling them both the ideal target and what is an acceptable compromise, if such a compromise is needed. This way the team and the owner can establish a more truly honest working relationship.

And fourth, it is much better for a team to know up front that an objective is either the ideal target or else the least you can accept, rather then for them to second guess this and have a 50% chance of being wrong, causing everyone a lot of frustration, wasted time and effort.

Some Additional Considerations

When developing project objectives there are additional considerations that can be important. **Figure 6** gives some of these considerations.

There can be times when project objectives have been given to the project leader and team, without due consideration to the material presented. In these situations the objectives will frequently be lacking in quality and need to be refined or re-done in order to be effective. There are times, in other words, when the project objectives really need to be re-negotiated.

- Verify that you are solving the right problem
- Talk to experts and those who might do the work
- Consider more than the first impression of what would be ideal for each of the triple constraints
- Are there better technologies or more cost-effective ones?
- What is changing in the market or new technology environment?
- How will key personnel affect this project?
- Document all the assumptions you are making
- Explore possible alternative solutions or objectives
- Prepare a draft set of objective statements, then test this draft with key people
- Refine the draft objectives and again review them with the key people
- Make sure that you have alignment of business and technical objectives
- Make sure that success metrics have been defined
- Assure that project implementation expectations are clear with all parties involved
- Once the objectives are finalized, constantly remind everyone involved of those objectives
- The longer the project duration, the more likely the project objectives might change or evolve—be alert for this possibility and manage the changes carefully through full communication

Figure 6. Some Additional Considerations for Developing Project Objectives

As Knutson and Bitz (1991) have stated well, this negotiation step "is a result of one of two sets of circumstances: (1) You approach the client and senior management for the first time in order to negotiate a schedule deadline and a budget because these were not established when the project was initiated, or, more commonly, (2) you realize that there is an incompatibility between the preliminary plan and the project objectives that cannot be eliminated by negotiating estimate revisions...In the latter case, the negotiations are more complex...There are four alternatives for management to consider:

1. Alter the schedule and cost objectives
2. Develop a reduced set of technical objectives (referred to as "Performance" objectives in this paper)
3. Implement a combination of the first two alternatives
4. Cancel the project

Regardless of the alternative you recommend and senior management decides on, there are only two acceptable outcomes to this process: (1) a plan acceptable to you, the project team, the client, and senior management or (2) cancellation of the project. There is one other possibility, however—one that we do not encourage: taking on the project on a best-efforts basis, without changing the project objectives."

If the procedures and tools suggested in this paper are followed and used, the above difficulties caused by poor project objective definition will be minimized or nearly eliminated altogether.

Once we have well defined, written goals or objectives, they need to be published for all people involved with the project. As Rees (1991) has said, "Publish the goals. Make them visible for all to see. Do not expect people to remember them if they are not discussed and referred to often. Do not bury them in occasional memos and documents. Instead, post them in meeting rooms on flip charts or posters; use stickers or other visual reminders to keep the goals in front of everyone. Refer to the goals often in memos, presentations, and meetings. Whenever possible, use the goals to guide a decision."

Finally, even after all the above work to do objective development right, to varying degrees the project objectives need to remain dynamic. How much the objectives are dynamic will depend on many things, and each project is unique. But the project objectives "should be reevaluated as the environment and opportunities change." (Rue and Byars, 2000) **Figure 7** gives some example factors that have the potential to cause objectives to change significantly.

- Market competition
- Significant changes in the national or corporate economy
- New senior management
- New discoveries
- New regulations
- Unknown technical difficulties in first-time scale-up to commercial production

Figure 7: Example Factors Potentially Changing Project Objectives Significantly

One important caution is appropriate here: When project objectives change very significantly, it is often much better to declare the old project dead and to officially start a new project with the new objectives. This is because the old baggage of plans and expectations from the original project can hinder the revised project to the point of making it inefficient and ineffective. It is very hard to change the direction of a speeding train locomotive.

Summary

In summary, defining good project objectives as early as possible is one of the most important project development steps to occur. Presented in this paper are a number of important tools and concepts to help project owners and project managers with this extremely important responsibility. Particularly we have discussed in detail SMART objectives, the triple constraints, the concept of prioritizing the project drivers, and the clarification that comes through use of Tolerance Diagrams.

All of these tools and concepts both help to facilitate and require much better multi-way communication between the project owner, project manager, project team members, and other key project personnel. The resulting improvement in project objectives definition will significantly raise the potential for project success.

References

Babcock, Daniel L., *Managing Engineering and Technology—An Introduction to Management for Engineers*, Prentice Hall (1991).

Bitz, Ira, *Senior Project Management*, Course Syllabus prepared for The American Management Association, revision number 3 (1991).

Dobson, Michael, from discussion at a Fred Pryor Seminar entitled *Project Management* (1993).

Frye, Chris and Holly Mason, "Project Management Simplified," Available from Internet: <http://www.webtechniques.com/archives/2001/09/frye/> (cited March 15, 2002)

Gibson, James L., John M. Ivancevich, and James H. Donnelly, Jr., *Organizations—Behavior, Structure, Processes*, Irwin/McGraw-Hill (1997).

Kepner, Charles H. and Benjamin B. Tregoe, *The New Rational Manager*, Princeton Research Press (1981).

Knutson, Joan and Ira Bitz, *Project Management—How to Plan and Manage Successful Projects*, AMACOM (1991).

Rees, Fran, *How to Lead Work Teams: Facilitation Skills*, Pfeiffer & Company (1991).

Rue, Leslie W. and Lloyd L. Byars, *Management—Skills and Application*, 9th ed., Irwin/McGraw-Hill (2000).

Smith, Larry A., *Basic Project Management*, from course materials reprinted with permission for use in American Management Association seminars (1991).

Overview of Project Planning

Paper # 474827
Dr. Eldon R. Larsen
Marshall University

Effective, excellent project planning is one of the vital three keys to successful project management, the other two vital keys being the importance of people and the importance of excellence in defining the project objectives.

When we say *planning* we are including *re-planning* as well. Re-planning is just a word we use to indicate the fact that the nature of planning should be ongoing continually throughout the life of the project. In fact, one very fine project manager I once met said that re-planning is really synonymous with controlling the project. Although people might say such a statement is over-simplified, in many ways this is a good way to think about the need to continually plan and re-plan.

To further illustrate an important point about planning, I use the following story. During the Persian Gulf War in the early 1990s the United States used some new, high-tech weaponry. Yet that alone was not enough to win the war, and it had only part to do with how the war was executed. In fact, many people recognized that the execution of the war was being successful in large measure due to some excellent leadership and decision being made by those in charge.

One day during the Persian Gulf War, at a news conference, a reporter asked the following question, recognizing how difficult it must be to plan and conduct a war when the enemy was continually moving around, hiding in difficult terrain, and moving scud missiles around on the back of trucks. The question was something to the effect: "General Schwarzkopf, how do you go about executing this war when the enemy is continually doing things that are not in your plan?" General Norman Schwarzkopf, quoting another famous general, Dwight D. Eisenhower, from World War Two, said in response that "Plans are worthless." Then, following a pause, continuing to quote, he said, "Planning is essential."

What did Generals Schwarzkopf and Eisenhower mean when they said that plans are worthless, planning is essential? In essence what they were saying is that plans are not sacred and they are not fixed! Many events and changes occur, new knowledge is learned, situations change, changes in leadership occur, competitors make unexpected decisions or announcements, people issues arise, and so on and so on and so on, all combining to make it completely unrealistic to think that somehow a plan can remain fixed! And yet there are a very large number of managers today who do not seem to understand these realities.

It is not *the plan* that adds value in project management! Rather, it is the planning *process* that adds value! It is through the *process of planning* that we have to think about all the various aspects of the project; we have to think about the various options and trade-offs; we have to think about what competitors might be doing; we have to think about alternative courses of action and the associated consequences; and we have to keep doing this throughout the entire project!

For a manager or anyone else to expect a project team to make a plan and then to religiously stick to it exactly with no adjustments is project tyranny. A manager who does this will cause his people to hate planning. Plans will become like strait-jackets, or like a noose around people's necks. This does not mean that the project objectives change or are flexible, although in some cases they might have to be, but rather it means that the way we achieve those objectives will continually have to be adjusted, modified, updated, and revised. Planning and re-planning are a way of life for project managers if they want to be successful. In fact, without re-planning, it is very likely that the project objectives will not be achieved! The bottom line to remember is this: In order to achieve project objectives, it is the continual *process* of planning that is essential and adds value, not the plan itself!

Throughout this presentation, we will be discussing many very important project management planning tools to help us in the process of project planning. These tools are very important and useful. Still it is not the tools themselves that are sacred, but the value of the tools comes through the process of using these tools. Use them appropriately.

The project planning tools in this presentation follow one another in a logical pattern. This pattern might be thought of as a cycle of project planning. The cycle is iterative.

When a project is first being planned, it might be necessary to iterate through the project planning cycle several times in order to get the initial plans and project objectives to be internally consistent with each other. During this initial iterative process through the planning cycle, an organization might find that it must make some adjustments to the original project objectives. This might happen when the planning cycle reveals better estimates for the project duration, project costs and other resource requirements, and the project performance expectations, and those estimates are not compatible with the project objectives. As adjustments are made in the objectives and the consequences of those adjustments are analyzed in the project planning cycle, an internally consistent plan and objectives can be achieved, or the project might need to be cancelled or significantly re-thought.

The project planning cycle generally consists of the following major steps:

→ Project charter development and elaboration
  - Project objectives definition
  - Identification of critical success factors
  - Definition of project standards
  - Identification of project leader and team
  - Elaboration of project deliverables
  - Other charter concerns

→ WBS development and elaboration

→ Network diagram development and elaboration

→ Estimates of resources, including costs, and durations for project tasks

- Note: Once this information is developed, a check for best-logic-consistency should be made in the network diagram

→ Development of resource loading information (resource histograms)

- Note: Once this information is developed, a check for best-logic-consistency should be made in the network diagram and the succeeding estimates step

→ Development of project cash flow, cumulative cash flow, and overall budget

→ Development and elaboration of plans for managing procurement, quality, risks, communications, scope, integration, human resources, time, and costs

→ Development and elaboration of contingency plans

- During and following development and elaboration of contingency plans, it is best to check for best-logic-consistency starting back from the WBS development through this step involving contingency plans, and all the steps in between (WBS, network, estimates, resource loading, cash flows and budget, and the other management process plans just before this contingency planning step)

→ Cycle back to check for internal consistency with the project charter

The above steps in the project planning cycle are shown graphically in Figure 1. After the initial iterations through this cycle which take place during the latter part of project initiation, the project team, its leader, and all key project stakeholders should have a good idea of the realistic expectations for the project. This does not mean that all will go as initially planned—it will not—but the team will now be prepared the best it can to move forward with project execution, with a much better chance of having a happy and successful project closeout for all parties involved.

Yet, the use of the project planning cycle has really only begun when project execution starts. The project manager must regularly check and monitor all calculation, assumptions, estimations, and so forth in the planning cycle. This is the heart of re-planning and the ability to control the final outcome of the project.

Summary

The project planning cycle contains a series of steps that build one upon another. Each step in the cycle is important and essential, and should not be skipped except in the most unusual circumstances, such as in a very small project that possibly does not merit a full use of these tools and processes. A wise project manager will give sincere consideration to the concepts discussed in this presentation. Doing so will help him to achieve true excellence in this, the third vital key to successful project management: Excellence in project planning and re-planning.

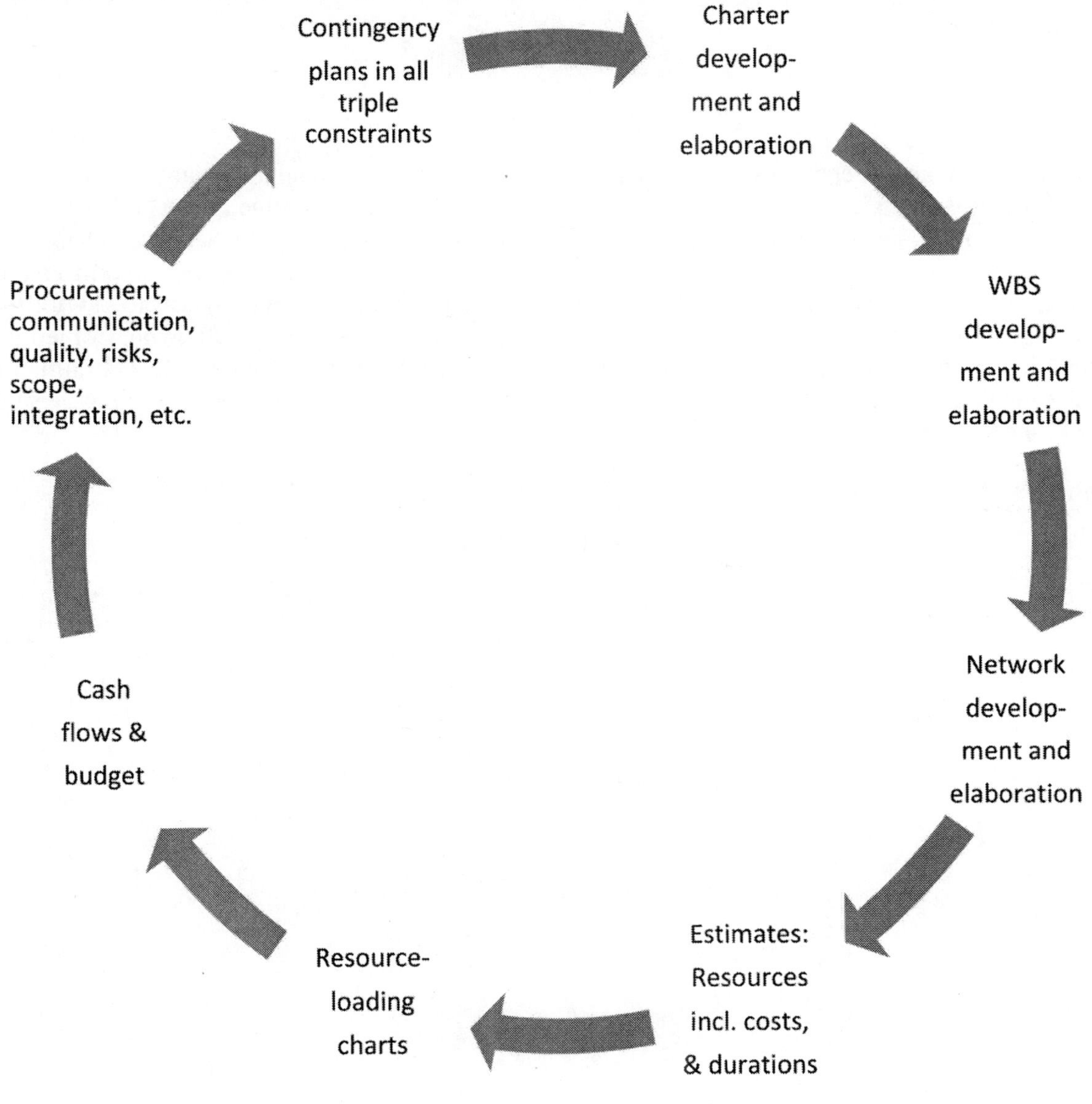

Figure 1. The Project Planning Cycle

481095 Design for Manufacturing " a New Approach to Engineering Design Education

Abstract:

Engineering design is often taught separately in each department as a problem with an artificially small number of design options and relatively simple economic models. While it exposes students to some aspects of ideation to generate design options and optimization to select the best one, it misses many critical real-world challenges. Firstly, design teams in industry are multidisciplinary, incorporating all of the engineering and business expertise required to develop the product. Secondly, the design should take into account the market and its willingness to pay for the product. Thus, the cost of manufacturing the product ultimately influences the product specification. Cornell is piloting a new approach to design education that allows students to traverse the full spectrum, from product definition to the manufacturing process that will deliver the product at spec and at the right price. We will present two case studies on this new approach to design education.

481094 Citrate-Based Macromolecules: A Versatile Biomaterial Platform for Regenerative Engineering

Abstract:

Advances in biomaterials science and engineering are crucial to solving complex medical problems that involve the replacement, repair, or regeneration of tissues and organs. Our laboratory pioneered the development and applications of citrate-based macromolecules (CBMs). CBMs can be engineered to exhibit physical, chemical, and biological functionalities that are suitable to solving a variety of medical problems and potentially treating disease conditions in the circulatory, musculoskeletal, renal, intergumentary (skin), and endocrine systems. I will discuss the properties of CBMs and fundamental design strategies that render them a powerful biomaterial platform for cell delivery, wound healing, and regenerative engineering applications.

481093 Regenerative Engineering: A Convergence of Enabling Technologies

Abstract:

We define Regenerative Engineering as a Convergence of Advanced Materials Science, Stem Cell Science, Physics, Developmental Biology, and Clinical Translation. Work in the area of musculoskeletal tissue regeneration has focused on a number of biomaterial technologies. Polymeric nanofiber systems create the prospect for biomimetics that recapitulate connective tissue ultrastructure allowing for the design of biomechanically functional matrices, or next generation matrices that create a niche for stem cell activity. Polymer and polymer-ceramic systems can be utilized for the regeneration of bone. Through the use of inducerons, small molecules fostering induction, the design of regeneration-inducing materials can be realized. Hybrid matrices possessing micro and nano architecture can create advantageous systems for regeneration, while the use of classic principles of materials science and engineering can lead to the development of three dimensional systems suitable for functional regeneration of tissues of the knee. Through convergence of a number of technologies, with advanced materials science playing an important role, we believe the prospect of engaging future grand challenges is possible.

477664 Effects of Hypoxic, Aligned Collagen Gels on Sarcoma Cell Proliferation and Invasion

Abstract:

Orientation of collagen fibers has proven to affect cell proliferation and adhesion to in vitro scaffold. More understanding of this effect on malignant cancers, like sarcoma, is desirable, especially under hypoxic conditions that many sarcoma tumor masses possess. Here, we evaluate the consequences of three-dimensional aligned collagen I fibers in extracellular matrix (ECM) on sarcoma cell proliferation and invasion in hypoxic regions. Applying a unidirectional compression on collagen gel has shown to increase the degree of fiber alignment, so that mimics the natural ECM orientation in tumors. By modulating the cell number encapsulated in gels, as well as the gel height, we can alter the oxygen diffusion rate and generate hypoxic regions within this in vitro model. To analyze sarcoma cell behavior, we culture individual sarcoma cells and small mice tumor grafts in hypoxic, aligned collagen gels and compare with their randomly oriented counterparts in atmospheric conditions. The topographical cues of

ECM, such as collagen architecture, influence the cancer cell proliferation and motility through altering cell-fiber interaction. From this we further study collagen modifiers/cross-linkers as well as focal adhesion kinase, as they influence cancer proliferation in the oxygen-extracellular matrix (ECM) tumor microenvironment. We show corresponding factors that correlate with the proliferative response of sarcoma cells in this truly physiologically mimicking platform.

477661 The Effect of Nanocatlyst Particle Size on Apparent Reaction Rates

Abstract:

Nanocatalysts serve in a number of chemical process applications for the petroleum and chemical industries. However, many aspects of the nanoscale that affect particle reactivity still remain unknown. Current mass transfer models do not extend into the nanoscale region and previous research findings conflict on the nanoparticle effect. The aim of this research project is to clarify and determine the effects of nanocatalyst particle size on the apparent reaction rate. A series of experiments were conducted with varying nanocatalyst particle sizes while holding catalyst surface area constant to prevent the confounding effects of increasing surface-to-volume ratio. Ferricyanate was reduced to ferrocyanate using gold nanocatalysts using UV-Vis spectrophotometry to track the progress of the reaction. Employing sodium borohydride as the reductant in sufficient concentration to establish a diffusion-limited reaction, the reaction rate increased with decreasing particle size as expected in view of the increased reactivity of surface gold atoms. However, the rate was orders of magnitude lower than that predicted by diffusion-limited theory. Using potassium iodide as the reductant, reportedly a kinetically-limited reaction, the reaction rate decreased with decreasing particle size. The iodide results are counterintuitive and even express the ability to reduce the rate below the uncatalyzed reaction. Clearly, the apparent reaction rate is influenced not only by particle size but by other factors, such as the intrinsic surface reactivity, stabilizing surface ligands, electrostatics, mass transfer and particle geometry. Future work is focused on understanding and ranking the effects of these various factors on the apparent reaction rate on nanocatalysts.

477655 Mathematical Modeling of Podocytes in Diabetic Kidney Disease

Abstract:

One of the major complications caused by diabetes is diabetic kidney disease (DKD), which is the leading cause of kidney failure. Despite the impact of this disease, the intricate details of how diabetes progresses to DKD are not well understood. Diabetes is characterized by elevated levels of glucose in the blood. The kidney is the main filtration unit of the body; as such any, severe adjustments to concentrations in the blood can affect kidney function. Filtration within the kidney occurs in the fundamental filtration units of the kidney; each of the thousands of filtration units is called a glomerulus. The death of podocyte cells that constitute the outer layer of a glomerulus begins a chain reaction of failure that ultimately leads to dysfunction of the kidney altogether. Due to the significant role of podocytes in the onset and progression of DKD, I have investigated the types of podocyte damage induced by diabetes that lead to such failures. The key question is how chemicals within the kidney alter podocyte function leading to DKD. I have utilized mathematical modeling to address this question by

formulating equations based on published data to test hypothesized mechanisms of the connections between different factors that affect the health of podocyte cells.

477663 Wearable Sensor Comprised of Nafion-Single-Walled Carbon Nanotubes-Metal Oxide Nanoparticles for the Detection of the Chemical Warfare Agent Simulant DMMP

ABSTRACT: The imminent threat of colorless and odorless chemical warfare agents (CWA), such as nerve agents containing organophosphorus, has spurred intense efforts towards the development of wearable gas sensors that can finely and selectively detect the presence of these lethal compounds. In this regard, breathable poly electrolyte membranes (PEM), such as Nafion, can serve a dual purpose. First, a semi-permeable membrane can be used to inhibit the diffusion of chemical warfare agents (CWA) while allowing for the diffusion of water arising from perspiration. Second, upon solvation, Nafion is capable of hosting the growth of semiconducting nanoparticles commonly used for the detection of CWA. Herein, a composite material comprised of both carbon nanotubes and zinc oxide nanoparticles was employed for its unique conductive and semiconductive properties, respectively. The resultant composite material is characterized by x-ray diffraction (XRD), Raman spectroscopy, and scanning electron microscopy/electron diffraction spectroscopy (SEM/EDS). Finally, silver metal contacts are evaporated onto the sides of the wearable composite, and the change in resistance is measured as a function of dimethyl methylphosphonate (DMMP) vapor concentration.

477657 Anhydride Synthesis through Reactive Distillation

Abstract:

Symmetric anhydrides are important chemical intermediates used primarily in Friedel-Crafts acylation reactions. In particular, the anhydrides of fatty acids (C12-C18) are biomass-derived molecules valuable for the synthesis of surfactants and other fine chemicals. There are stoichiometric processes to prepare these anhydrides in the laboratory but a catalytic, continuous process does not exist at this time. We have investigated the use of reactive distillation to drive to completion the exchange reaction depicted below:

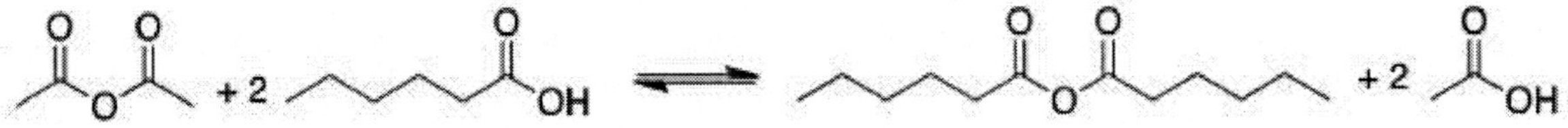

A batch reactive distillation setup was tested as a viable and efficient way of making long chain anhydrides from acetic anhydride and the corresponding acid by continuously distilling the acetic acid as it is formed. Here hexanoic acid was used as a model fatty acid. Initially, we determined whether a catalyst is needed to achieve a high conversion of acid. Zeolites, acid resins, and a homogeneous acid catalyst were used to catalyze the reaction of hexanoic acid and acetic anhydride to form hexanoic anhydride. It was found that the reaction is able to proceed to the same extent and at comparable rates both catalyzed and uncatalyzed, and thus a catalyst is not necessary. In the uncatalyzed system, a high conversion of both the hexanoic acid (91%) and the acetic anhydride (83%) were observed after 2.5 hours. However, at this level of conversion of hexanoic acid, the mixed anhydride was produced instead of the desired symmetric anhydride (yield of 25%.) When the reaction was run for only 1.5 hours, a 60% conversion of hexanoic acid was seen with a 22% yield of hexanoic anhydride. Additional experiments will be conducted to maximize the yield of symmetric anhydride and larger alkanoic acids will be also investigated. The batch distillation results will be used to design and simulate, using Aspen+, an industrial scale reactive distillation unit to produce long chain anhydrides.

477662 Sonification of Bacterial Chemotaxis: Listening for Patterns in Response of Swimming Bateria to Chemical Stimuli

Abstract:

Chemotaxis describes the movement of a bacterial population toward or away from chemical stimuli that will enhance their chance of survival. Bacteria such as *Escherichia coli* and *Pseudomonas putida* commonly exhibit this behavior. *E. coli* and *P. putida* swim about the environment through the use of flagella, tail-like appendages that protrude out of the cell. Chemotaxis is an important phenomenon to understand due to its role in the development of infection and processes such as nitrogen fixation. Sonification is the use of sound to represent data and allows one to also hear the data rather than just see it graphically. The human ear is trained to detect patterns and slight differences in sound. Thus, sonification may be an effective way to observe a bacterial populations response to chemical stimuli. Sonification of data can shed new light on relationships and patterns in the data that may not have been realized otherwise. Time lapse images using dark field microscopy were taken of HCB1 *E. coli*, *P. putida* F1, and a mixed population of both HCB1 *E. coli* and *P. putida* F1. The bacterial responses to varying concentrations of attractants and repellants were captured. These time lapse images were then sonified by mapping position to frequency of sound. Sound maps of the data reflected the presence of varying concentrations of stimuli and the change in swimming patterns of the bacteria. These preliminary results demonstrate sonification as a method to observe chemotactic responses and may be used to study bacterial responses to a wide array of chemical stimuli.

478912 Investigation of Functionalized Spherical Silica As a Hydrate Blockageinhibitor in Water-in- Oil Emulsions Under Quiescent and Flowing Conditions

Abstract:

Hydrate mitigation and management is a growing area of flow assurance engineering with crystalline structures forming in water-in- oil emulsions under low temperatures and high pressures often leading to pipeline blockages. Ludox TM-50 and powdered spherical silica particles will be surface treated to hydrophobize the particles for use as a stabilizer in water-in- oil emulsions to investigate their effect on hydrate formation. The first analysis of hydrate formation in water-in- oil emulsions stabilized using silica particles will be using optical microscopy and particle distribution characterization techniques. Experiments will test the effect of various water cuts, temperatures, and concentration of solid particles.

After a fundamental understanding of the behavior of silica particle stabilization is reached, the focus will shift to analyzing water-in- oil emulsion behavior under flowing conditions using a flow loop system. These experiments are expected to determine the utility of the Ludox TM-50 and powdered silicon dioxide particles as a method of hydrate inhibition and understand their effect on hydrate formation under both quiescent and flowing conditions. Work on this project is ongoing.

477658 Effects of Nanoscale Geographies on Osseointegration and Bacterial Growth

Abstract:

Titanium has been widely used in bone implants, owing to its near-ideal mechanical strength and relative biocompatibility, but often necessitates repeated surgeries due to infection and poor osseointegration. Implants can be roughened with porous coatings to encourage cell integration, and with nanoscale titanium texturing, there have been results of improved osseointegration and decreased bacterial presence.

It is known that nanoscale features improve bone growth and reduce bacterial adherence, but unknown exactly which structure of this texturing is most conducive to this. From previous bacterial experiments with the titanium samples, there appeared to be no strong correlation between the colony forming units of seeded S. aureus and the sample roughness, which suggested that the current method of quantifying roughness with a specific value may be flawed. Comparison of atomic force micrographs of samples alongside their â€œroughnessâ€ values

evidenced that a specific value did not imply a particular nanotexturing; some features displayed similar roughnesses yet possessed dissimilar surface features. Investigation of these nanogeographical differences demonstrated that samples with more dense, evenly sized nanosurfaces resulted in greater bacterial inhibition than those with more varied nanofeatures. Regardless, that all nanostructured samples exhibited strong antibacterial properties with little detrimental effect on human osteoblasts.

477659 Modeling Anaerobic Digestion of Biodegradable Solid Waste

Abstract:

Anaerobic digestion (AD) is the biological conversion organic matter into methane gas. The US Air Force aims to use this process to treat the biodegradable solid waste produced on their forward operating bases (FOBs). The focus of this study was to obtain the Monod kinetic parameters of AD process intermediates (volatile fatty acids, monomers, and polymers) that are typical of waste produced on FOBs. To obtain Monod kinetic parameters a combination of experimental and computational methodologies were applied. First, the AD of a single concentration of each of the substrate was regressed to obtain preliminary saturation constants (K_s) and maximum rates (μ_{max}) and next, these preliminary values were used to obtain actual experimental values. The experimental values obtained for the volatile fatty acids include: K_s = 0.28 g L^{-1}, μ_{max} = 0.00080 g COD g VSS^{-1} h^{-1} for acetic acid, K_s = 0.22 g L^{-1}, μ_{max} = 0.00049 g COD g VSS^{-1} h^{-1} for butyric acid, and K_s = 0.089 g L^{-1}, μ_{max} = 0.00031 g COD g VSS^{-1} h^{-1} for propionic acid. The parameter values obtained were within the range of previously reported

literature values. Once all of the kinetic parameters are obtained, the parameters will be incorporated into the Anaerobic Digestion Model No.1 (ADM-1). ADM-1 predictions will be compared to batch AD performance for food waste, and model deficiencies will be discussed.

477666 Mosaic Ion-Exchange Resins for Reducing the Overpotential for Water Dissociation in Bipolar Junctions for Electodeionization Separations

Abstract:

Electrodeionization (EDI) is an electrochemical separation technology capable of removing salt from water using an applied electric field, ion-exchange membranes, and ion exchange resin bed in the diluent stream. [1] The porous ion-exchange resin bed reduces the purified stream resistance, the largest resistance in the system, by conducting ions across the immobilized bed and by dissociating water to generate H+ and OH- ions for current flow through the stack. Water dissociation in the ion-exchange resin bed occurs in nanometer thin bipolar junctions in the resin bed. [2-4] These junctions consist of oppositely fixed positive charges that generate an internal electric field for dissociating water. Most ion-exchange resin beds for EDI employ commercially available cation and anion ion-exchange resins mixed together, but such mixing of the different beads results in underutilized resins because few fixed anionic and cationic groups form effective

bipolar junctions. The consequence of poorly formed bipolar junctions results in energy inefficient water desalination via EDI.

Page 2

In this work, mixtures of thin cation and anion polymer electrolyte brushes were thermally grafted to silicon wafer and fumed silica nanoparticle surfaces to realize thin film mosaic ion-exchange materials for EDI. Grafting the oppositely charged ion conductors to the same substrate leads to a greater probability of opposite fixed charges being in close proximity for water dissociation. The process to prepare the mosaic materials started with thermally grafting mono-hydroxy terminated non-ionic polymer brushes to the substrate via a condensation reaction under an inert environment. Then, subsequent ionization reactions (e.g., Menshutkin, acetyl sulfonation, or thermal decomposition/acid catalyzed hydrolysis) [5-7] converted the brushes into anionic and cationic polymer electrolytes. Verification of brush grafting and introduction of ionic moieties was made using water contact angle measurement and Fourier transform infrared spectroscopy (FTIR). Ionic conductivity of the homo- and mixed-polymer electrolyte brushes to interdigitated electrodes was conducted via electrochemical impedance spectroscopy. Future studies aim to incorporate the mosaic materials for resin-wafer EDI.

474593 Innovations at the Nexus of Food, Energy, and Water Systems (INFEWS): A National Science Foundation (NSF) Initiative to Examine Critical Issues in Today's World

Abstract:

At the end of 2015, the National Science Foundation (NSF) put out its first request for proposals for the INFEWS program (Innovations at the Nexus of Food, Energy, and Water Systems). Over the next five years the program plans to fund interdisciplinary projects that fall into four tracks 1. System Modeling, 2. Visualization and Decision Support, 3. Innovative System Solutions, and 4. Education and Workforce Development. The program expects to tackle Food-Energy-Water System (FEWS) research through integration across disciplines, in which agricultural and food systems research is a primary component. To this end NSF has partnered with the United States Department of Agriculture (USDA) to fund related research. This program developed from the Science, Engineering and Education for Sustainability (SEES) program, which in 2015 funded 17 FEWS workshops. These interdisciplinary workshops brought together researchers and other stakeholders from universities, local, state, and federal governments, industry, and non-profits to examine a wide range of FEW questions and identify the research, data and technological needs to solve them. This presentation will cover an overview of the development of the INFEWS program, some pressing FEW needs as identified by the broader research community, and national and international opportunities for future engagement with the program.

474707 The "Root" of the Nexus: Soil-Based Biotechnology for Sustainable Agriculture

Abstract:

Food, energy, and water nexus research by chemical engineers is typically focused on the built environment. For example, innovations in sustainable food production might offer food processing, distribution, or waste recycling technology that minimizes water or energy use. However, recent technological advances in different fields ranging from materials and manufacturing to molecular tools and systems biology are also enabling more sustainable primary production of food in the natural environment. Most of our food is produced in soil, a dynamic natural system of enormous complexity. In particular, the rhizosphere (the region of soil adjacent to plant roots) is a site of steep chemical gradients, plant growth-limiting multiphase transport phenomena, and dynamic inter-kingdom collaboration among members of the root microbiome. Emerging technology is re-engineering the root zone for more sustainable primary food production. Examples of emerging agriculture biotechnology to produce more food with less water and energy include: (i) advanced seed coating technology and nano-enabled agrochemical delivery systems; (ii) plant-growth promoting bacteria and microbial products to improve plant nutrition and enhance moisture retention in soil; and (iii) sensors and sensor networks for precision agriculture. By learning from and working with natural systems, chemical engineers are developing innovative agriculture biotechnology for more sustainable primary food production.

474596 The Role of Waste Streams in the Energy-Water Nexus

Abstract:

The energy-water nexus has helped focus attention on the water used for energy production and energy used in providing water services. However, there has been a lack of attention to the downstream aspects of this nexus; that is, impacts of related waste streams on land, water and air. Primary technology choices (e.g., electricity generation technology, drinking water source) result in differing waste streams which in turn require decisions on the level of treatment and disposition of the related waste products. Each of these choices have related implications for energy use, water use, environmental impact and system vulnerability (e.g., power plant thermal discharge puts operations at risk to drought). Remarks in this panel will endeavor to establish a framework and context for a more holistic treatment of waste streams within the broader energy-water nexus.

474597 Process Development Challenges and Opportunities in Food-Energy-Water (FEW) Nexus

Abstract:

Presentation will start out with a high level view of Food-Energy-Water (FEW) nexus for illustrating complexity of the topic. Global changes in population, energy needs, and water use will drive technological innovations and rapid commercialization needing chemical engineers involved in process development. Connections and interdependencies discussion among these three systems will illustrate how engineering solutions from fields like systems approach, catalysis, biocatalysis, environmental sustainability, etc. will contribute to better understanding the FEW nexus challenges and successfully overcoming them.

Broad overview of current and future process technologies important to FEW nexus selected from various NSF and AIChE workshops and recent developments reported in literature will be presented. Challenges and opportunities for successful transition from concepts to commercialization of such technologies will be discussed with actual examples at various stages of development. Examples will be presented wherever possible to show how process research and innovation, pilot plants, technology transfer, process intensification and micro process engineering as well as product design will contribute to the nexus. Need for collaborative efforts among academics, government labs, and industry across multiple disciplines and countries for local applications for ultimate success in the FEW nexus will be highlighted.

The talk will also cover selected areas where chemical engineers in general and process development engineers in particular can make significant contributions. Critical need for process research, development, and demonstrations that enables new means of adapting to future challenges will be presented. As an example, out of the three main factors nitrogen, phosphorous, and water that limit our ability to produce enough food, nitrogen and water will be covered in detail. This section of the presentation will show how catalytic and waste water treatment technologies will play the role in connection to the FEW nexus.

Main goal of the presentation will be to make audience aware of theoretical and experimental process development challenges and opportunities of the future.

474595 Systems Engineering Methods for Food-Water-Energy Nexus

Abstract:

As the global population grows, consumption of water, energy, and food will also increase, placing stresses on these three sectors, raising the importance of the Food-Water-Energy Nexus (FWEN). This article highlights research challenges and identifies process systems engineering research opportunities to appropriately model and optimize the WEFN. A brief overview of relevant, foundational FWEN research is first presented. We then identify challenges in the multiple scales of the FWEN, ranging from the household scale to the global scale. There are further challenges with appropriate system boundary definitions, and challenges in modeling the decision-making and conflicting objectives of multiple stakeholders in the FWEN. Uncertainties of all kinds appear at all scales of the FWEN and must also be considered. We use two motivating FWEN examples to frame these challenges and propose future avenues and opportunities for research. Possible approaches to the aforementioned challenges are proposed.

473481 Retractions, Post-Publication Peer Review, and Fraud: Scientific Publishings Wild West

Abstract:

Beginning in 2010, a database documenting retractions in the scientific literature was developed. Since that time, retractions have nearly doubled. This presentation discusses probable reasons for the increase and the ramifications of erroneous data in the literature. The website describing the retractions has over 12,000 subscribers and receives over 500,000 views per month. Philanthropies, including the MacArthur Foundation, support this project.

477618 Destruction of Chlorofluorocarbons in Non-Thermal Plasma

Abstract:

Non-thermal plasma has been studied as a prospective technology for remediation of gas streams contaminated with VOCs, NOx, SOx, and other compounds. The research is directed towards studying the mechanisms of destruction of chlorofluorocarbons (CFCs), specifically CH2ClF and CCl3F in a corona discharge - generated non-thermal plasma. The destruction of CFCs was studied in a tubular dielectric barrier AC corona discharge reactor coupled to an on-line quadrupole mass-spectrometer. The spectrum and concentrations of the products of the destruction were measured at the outlet of the reactor. The efficiency of the destruction was studied as a function of the active power dissipated in the reactor and the flow rate. The research showed that CH2ClF and CCl3F can be effectively destroyed by corona discharges in non-thermal plasma at a concentration of 1000 ppm. The results confirm the theory that destruction efficiencies are higher for compounds that have lower ionization energies than molecular oxygen. The ionization energy of molecular oxygen is 12.07 eV, and those of CH2ClF and CCl3F are 11.71 eV and 11.68 eV respectively. Lower concentration studies will be performed in the future, to test if the destruction efficiency increases as the CFC concentration decreases. CFCs are known as ozone depletion agents via the so-called chlorine catalytic cycle. The research directly applies to the possible methods for utilization of CFC stockpiles and has applications to atmospheric chemistry models.

The Lumburnator, A Next Generation Wood Stove

Student Team:

Jake Lindberg, Kevin Lee, Yunxiao Jiang, Dongni Qiu, Angell Chee, Jason Peng

Advisors:

Dr. Devinder Mahajan,
Dr. Tom Butcher,
Dr. Jon Longtin,
Dr. Vladimir Zaitsev

NYSERDA Grant Manager: Robert Carver

Introduction:

- Wood stoves are used to heat many people's homes, but what makes one better than the others?
- A desirable wood stove will find a balance between achieving high heating efficiency and emitting less carbon monoxide (CO), particulate matter (PM), and other pollutants.
- Our research looks at the efficiency and emissions of the wood stove when we:
    - Add an afterburner system
    - Use a Venturi draft inducer in the flue stack
    - Add Baffles to the combustion chamber

Afterburner: Concept

- Secondary combustion is used to reduces emissions from wood stove.
- Secondary combustion can be performed using a catalyst or afterburner.
- We use a propane-fueled afterburner to achieve secondary combustion of exhaust gases.
- Afterburner can perform secondary combustion during startup while a catalyst can not.

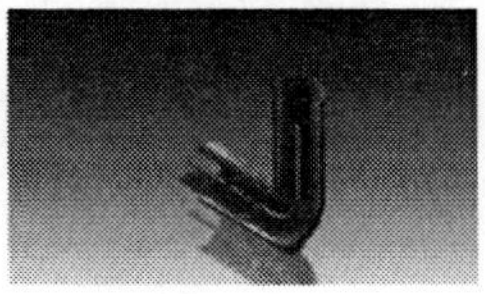

Afterburner: Proof of Concept Results

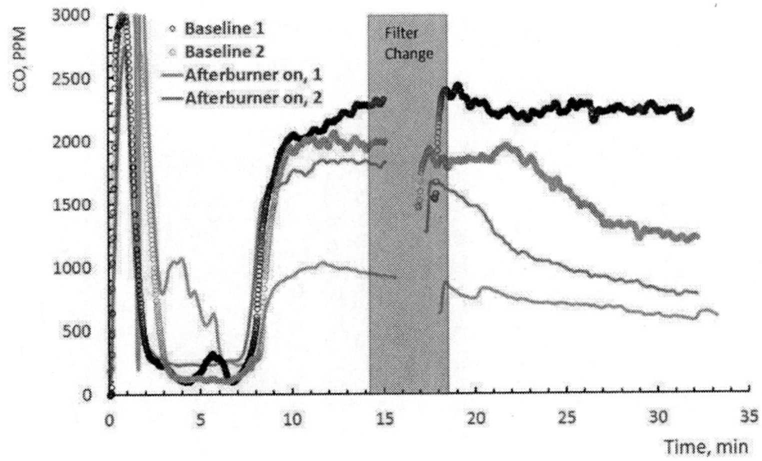

- Afterburner unit could reduce CO concentrations
- Performance will be tested
- Viability Factors:

1) Rate of Propane consumption
2) Afterburner's CO production
3) Afterburner safety

Air Flow System (AFS)
Venturi Draft Inducer: Concept

- Induced draft in the flue stack
- Extra oxygen to assist combustion
- Regulate burn rate
- Dilute exhaust, lower the temperature

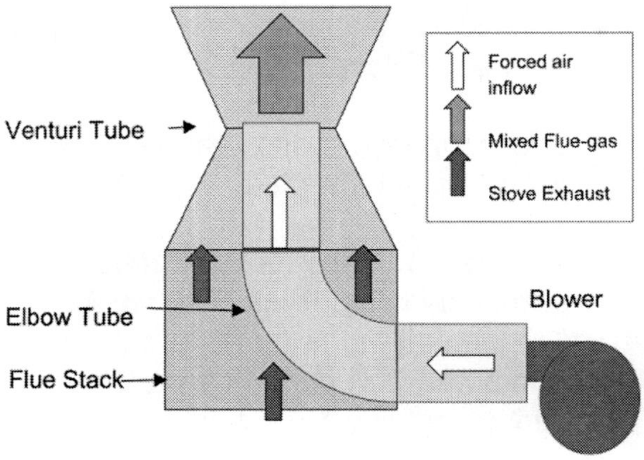

Dividers: Concept

- Increase gas residence time
- Increase turbulence in the firebox
- Create staged combustion
- Divide chamber volume 1:1

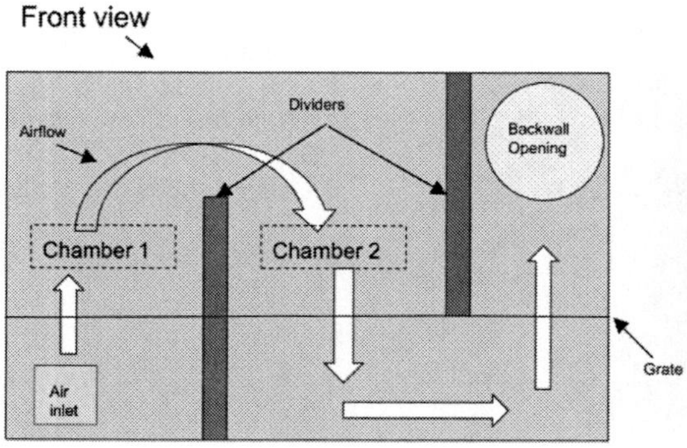

Preliminary Performance Data : Efficiency

NEW YORK STATE OF OPPORTUNITY. | NYSERDA

- Heating Efficiencies are different from ~60 min. to 90 min. into burn.
- Staged combustion improves heating efficiency.
- AFS extends period of heating efficiency.

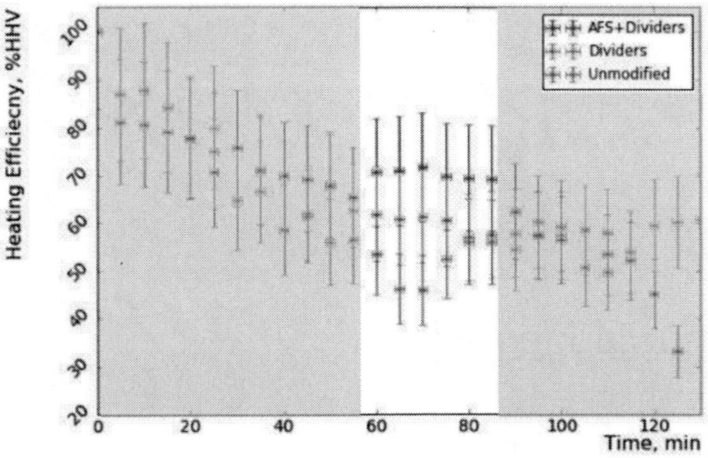

Preliminary Performance Data:

NEW YORK STATE OF OPPORTUNITY. | NYSERDA

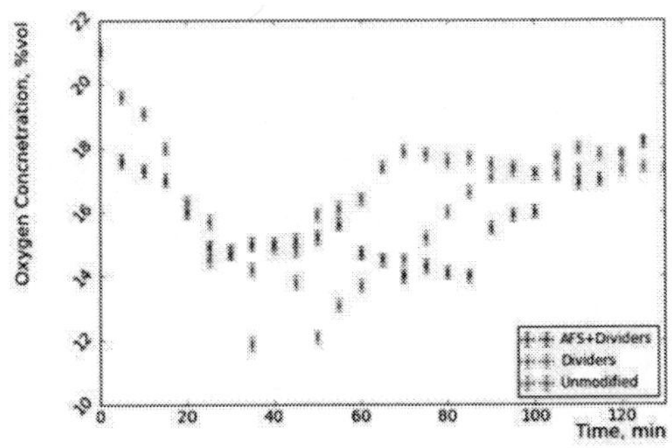

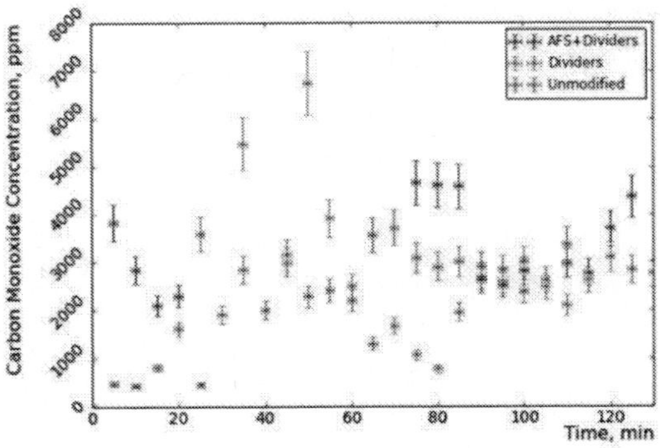

- Test results: Wood stove unmodified and with 2 modifications (with dividers, and with AFS and dividers)
- Modifications to wood stove create more CO
- Not enough O_2 in modified firebox

Acknowledgement

Funding source:
New York State Energy Research and Development Authority (NYSERDA) under Agreement No. 63043).

Special thanks to:
Robert Carver, NYSERDA
Dr. Tom Butcher, Brookhaven National Laboratory
Stony Brook University Professors:
Devinder Mahajan
Vladimir Zaitsev
Sotirios Mamalis
Benjamin Lawler
Jon Longtin

477619 Quantification of the Effect of Solids in Solution on Gas Evolution Rates

Abstract:

Gas evolution is an important phenomenon for various processes in the oil and gas industry, namely gas/liquid phase separation. Rates of gas evolution are a key uncertainty in gas/liquid separator design. Work is being conducted to gain a comprehensive understanding of gas evolution kinetics; however, the presence of solids in solution was previously unaccounted for. The purpose of this project is to develop a simple methodology to quantify the effect of the addition of solid particulates of varying characteristics (hydrophobicity, density, mass, and surface roughness) to gas evolution behavior of a supersaturated solution. This study details the design and commissioning of a cost effective gas evolution experiment with the ability to determine a gas evolution rate of a system with added solids at ambient pressure. An experimental setup using club soda as a supersaturated solution employed the displacement of a liquid to measure evolved gas and calculate gas evolution rates. Experimental trials have shown that a gas evolution rate increases with increasing mass of a dense hydrophilic solid introduced to a supersaturated solution. Work on the project is still ongoing.

477620 Composite Aminosilica/Polymer Hollow Fiber Sorbents for CO2 Capture from Flue Gas

Abstract:

PAPER TITLE: **Composite Aminosilica/Polymer Hollow Fiber Sorbents for CO_2 Capture from Flue Gas**

DATE SUBMITTED: 7/11/2016

ABSTRACT: Amine-loaded silica/poly(amide-imide) hollow fiber sorbents are created and used for CO_2 capture under simulated post-combustion flue gas conditions. Amine is infused into the mesoporous silica/ poly(amide-imide) hollow fiber sorbents during fiber solvent exchange steps after fiber spinning. The amine loaded fibers are tested by exposure to simulated flue gas at 1 atm and 35 °C. The amine functionalized silica/poly(amide-imide) hollow fibers comprising 5 wt% silica exhibit a high CO_2 equilibrium capacity of 1.85 mmol/g-fiber which is significantly higher than previously reported for fiber sorbents.

477617 Kinetics and Reactor Performance of Liquid Phase Ethylene Production By Hydrogenation of Acetylene

Abstract:

Ethylene, a major chemical component in the production of polyethylene, is primarily produced from the thermal cracking of naphtha with acetylene as the byproduct. The concentration of acetylene in the cracking effluent must be less than 5 ppm to improve the feed for polyethylene production. The gas phase hydrogenation of acetylene is currently the predominant method used in industry for removing acetylene from the cracking effluent. In this process, the formation of green oil on the catalyst surface contributes to its deactivation. However, liquid phase hydrogenation of acetylene has been gaining attention because it prevents runaway conditions caused by the exothermic nature of the reaction while also reducing the green oil formation on the catalyst surface. The focus of this work is to produce ethylene by hydrogenating acetylene in liquid phase using palladium catalysts supported on alumina. A polar solvent, N-methyl-2-pyrrolidone (NMP), was used to selectively

absorb acetylene and further hydrogenate to ethylene. This study will be the first in open literature to examine the kinetics of the liquid phase hydrogenation of acetylene using a selective solvent. The kinetics will be investigated using a basket reactor at a temperature and pressure range of 80 - 120°C and 200 - 250 psig, respectively. The performance of the fixed bed reactor will be studied under various operating conditions for upflow and trickle modes. The comparison between these reactors will be investigated and evaluated.

477616 Direct Synthesis of Hexagonal Boron Nitride on SiO2/Si Surfaces Via Chemical Vapor Deposition

Abstract:

The experimental findings of graphene in 2004 and its emerging impact in science and technology has open the avenues for other two-dimensional (2D) materials including hexagonal boron nitride (h-BN), transition metal dichalcogenides, black phosphorous etc. Traditionally, semiconducting 2D materials are being placed on SiO_2/Si surfaces for the measurement of electron transport phenomena. However, the surface roughness and charged impurity associated with SiO_2/Si surfaces reduce the nanoelectronics performance in interfaced 2D materials. In contrast, h-BN, which is a wide bandgap semiconductor with excellent properties such as: atomically smooth, lack of dangling bonds, high chemical and thermal stability and can be used as a gate dielectric for 2D nanomaterials electronics. Currently, there is no direct-growth mechanism for h-BN and most of the research rely on synthesizing h-BN on metal films and subsequent transfer for device applications. Considering the important role of ultra-smooth h-BN

in enhancing the carrier mobility of interfaced 2D materials, it is critical to grow large-area h-BN directly on desired substrates. Here we show high quality, large-scale and uniform h-BN films synthesized directly on SiO_2/Si substrates via surface-interaction guided chemical vapor deposition (CVD) as confirmed by combined spectroscopic and microscopic analysis. The weak Raman intensity at â‰ˆ 1373 - 1375 cm^{-1} corresponds to the atomically thin h-BN films. We also observe that Raman E_{2g} peak position shifts to higher frequency which may be attributed to the substrate induced stress in the films. Further, the effects of Si crystallography and CVD cooling conditions on h-BN formations will be discussed. This process of achieving h-BN avoids transfer related steps for further electrical characterization and integration into semiconductor processing.

477615 Effect of Biodiesel in the Enhancement of Gtl Diesel Fuel Properties

Abstract:

ABSTRACT:

Gas-to-liquid (GTL) fuels are environmentally friendly fuels due to their lower emissions of carbon monoxide and particulates upon combustion. Sulfur and aromatic content of the diesel fuel is detrimental to some of its physical properties such as lubricity and density and upon their absence in GTL fuels these properties are negatively impacted. This fact makes these fuels incompatible for use in existing diesel engines. Especially the lubricity possessed by the fuel which is significantly decreased for GTL diesel fuels, down to a level not meeting the minimum criteria for global standards. The aim of this research project is to modify the ultra-clean GTL diesel fuel to make it compliant with global standards and also to develop comprehensive knowledge to allow the possibility of stretching and tailoring the GTL diesel properties such that they meet the standards needed to be used without altering existing engines. The enhancement of lubricity, the most negatively impacted property, needs to be done while maintaining all other physical properties within the range needed for existing engines and American Society for Testing Materials (ASTM) standards. For this study, four lubricity enhancement biodiesel additives were selected. Blends were tested for fuel characteristics and physical properties. The tests include vapor pressure, viscosity, flash point, pour point, cloud point, distillation performance, heat content and lubricity. The work will be further continued to validate experimental results with partners in the oil and gas Sector to further improve the blends.

477614 Analysis of Mechanism of Nafion Conductivity Change Due to Hot Pressing Treatment

Abstract:

Proton Exchange Membrane Fuel Cells (PEMFC) are an emerging alternative energy source that provide high power density with only heat and water as waste products. A current challenge of making PEMFC technology broadly available is the performance losses and the cost of the membrane electrode assembly (MEA). The group previously observed that performing repeated hot pressing treatments on the PEMFC membrane, Nafion® 212, significantly increased the performance of the cell. The mechanism of these performance improvements was investigated by using various analytical techniques on samples subjected to different hot pressing treatments. The effect of hot presses on proton conductivity was investigated with Electrochemical Impedance Spectroscopy (EIS) measurements, which was used to separate and quantify the different sources of polarization for the fuel cell. To determine if the increased performance was a result of a chemical change, Thermogravimetric Analysis (TGA) and Fourier Transform Infrared Spectroscopy (FTIR) were performed. Results indicate that hot pressing the

membrane significantly affected the electrical properties of Nafion® 212, illustrated through the decreased series resistance in EIS measurements. However, TGA and FTIR measurements demonstrated that the changes in Nafion® 212 were not chemical in nature. This study suggests that hot pressing Nafion® 212 at the glass transition temperature causes an increase in proton conductivity due to water channel reconfiguration in the membrane.

477612 Surface Aging Effects of Primer Undercoats Used in Aircraft

Abstract:

UNDERGRADUATE RESEARCH FORM
AIChE 2016 ANNUAL STUDENT CONFERENCE
Sunday, November 13th and Monday November 14th
San Francisco, CA

***Please note there are three sessions, Sunday: 3:30-6PM and Monday: 12:30-3pm and 3:15-5:45pm. If you submit an abstract and are not able to present at a certain time, please let us know.**

NAME:	Austin Habich
UNIVERSITY:	University of Washington
EMAIL:	Austin.habich1@gmail.com
PHONE NUMBER:	509 670 7853
STREET ADDRESS:	12028 31st AVE NE, APT 304
CITY:	SEATTLE
STATE:	WA
ZIP CODE:	98125

PAPER TITLE:	Surface Aging Effects of Primer Undercoats Used in Aircraft
DATE SUBMITTED:	JULY 8TH 2016

ABSTRACT: Aircraft paints adhere well only with the aid of a primer undercoat, which ages after application thus requiring the primer to be applied within a time limit. It is now planned to carry out these operations in different locations around the world under different ambient conditions. Contact angle and peel adhesion tests are used to test surface aging under different conditions.

477613 Mixtures of Ultra-Low Viscosity Base Fluids for Fuel Efficiency

Abstract:

Tribology consists of the study of friction, wear and lubrication which is relevant when surfaces come into contact in relative motion. With oil, lubrication is implemented and the amount of wear and friction is greatly reduced. The goal of this project was to investigate low-viscosity base stock mixtures in order to improve fuel efficiency. A synthetic base oil, an ester, and an advanced mineral oil were selected to be tested with 5 different concentrations of each. By understanding the way these mixtures affect friction, wear, and overall lubrication of the metal to metal contact, we can better understand what would work best to increase fuel efficiency in an engine. A High Frequency Reciprocating Rig (HFRR) and a 4-Ball Machine were used to conduct the experiments since they imitate a Piston and Cylinder in a motor engine. A Light Optical Microscope and a Profilometer were used to analyze the wear scars and tribological films formed on the metal surfaces. The data was then organized and presented by using Excel, Kaleidagraph, and PowerPoint

466258 Innovations, Interactions and Integrations in Fluid Particle Systems; Memories and Future Tasks

Abstract:

Innovations, Interactions and Integrations in Fluid Particle Systems

Memories and Future Tasks

Masayuki Horio

Tokyo University of Agriculture and Technology

masa06@cc.tuat.ac.jp

1. Introduction

The early history of fluidization was inseparable from the huge innovation of the catalytic cracking technology. Such a tradition has been active until recently and we have been committing quite a few challenging developments in environment, energy, materials and pharmaceutical fields. As discussed elsewhere [1, 2], researchers of the fluid-particle engineering field were entering the second 30 years' period of scientific breakthrough at the time of early 1970s. During this second 30 years period our peer engineers have developed several commercialized and some partially commercialized processes and a few candidate processes that could be deployed in the coming future. However, shapes of industries are expected to change drastically for year 2050 and further from both necessities for environmental sustainability by mitigating global warming and for economic sustainability by developing a compromise among global, national and local economies. Also important is the sustainability of fluid particle education to keep our knowledge and technology active, interesting and renewable. Or in other words, fundamental scientific and engineering knowledge on fluid particle systems need to become more popular so that our classical achievements in the technical/engineering field can be inherited safely for long. Therefore, it would be worth discussing the shape of industries in the future and the shape of academia to prepare for a big transformation of the present social energy metabolism and the transition to the low-carbon and humanistic society. In this respect it should be stressed that innovations in the near future are becoming not just technical but also social. Both engineers and academic researchers would need more knowledge on social issues in the near future.

The era started early in the 1970s, which was the time I enjoyed most with my students and with many friend researchers in the world, was indeed a critical period of fluidization and fluid particle engineering in terms of innovations, interactions and integrations, but in several different senses as indicated above.

Concerning interactions, we have been concerned about several different categories of interactions: fluid-particle interactions, particle-particle interactions, hydrodynamics-plant interactions, chemical reaction-hydrodynamic interactions, and even the interactions among different branches of academia and between industries and academia.

Concerning integrations, we have achieved the integration of previously separated modes of fluid-particle systems including fixed/moving bed, bubbling fluidized bed, turbulent and fast fluidization, entrained bed and pneumatic transport, the integration of microscopic phenomena into mesoscopic and macroscopic ones and integrations of different branches of fluid-particle reaction engineering.

In the present paper let me take some memorable experiences of mine and discuss the three 'I's by leaving my thoughts.

2. Interactions — Incident of L-valves

When the license of multi-solid FBC boiler was imported and scaled-up by Mitsui Engineering & Shipbuilding (MES) from Battelle Columbus Laboratories, one of the most difficult issues was L-valve control. The original picture of the L-valve was beautiful in recycling solids from the external heat exchanger either by cold recycle or hot recycle to control the combustor temperature as desired. However, L-valves for the hot recycle were quite like a group of unruly children flowing free even in the no-aeration conditions. At that time the plant was waiting for the final transfer to the customer at the site. While discussing with an engineer from MES a suspicion came to my mind that some gas generation took place by chemical reactions. We started making a rough evaluation of CO formation and found out it was sufficient to fluidize the stand pipe.

There I learned that interactions between hydrodynamics and chemical reactions are two-way interactions, not just one way from hydrodynamics to chemistry, and chemical reactions to hydrodynamics could be critical particularly in high temperature processes. But from this incidence I was also convinced that any business risks have chances to be tightly related to some very fundamental phenomena and that relevant scientific knowledge has a potential to save them. Moreover, I re-acknowledged the importance of industry academia collaboration for the progress of technologies as well as engineering sciences.

3. Integrations — Two topics

1) DEM simulation

It is a talent required for researchers to look at phenomena always as integrated things. So, it is their common habit to decompose things into elements or essential laws and to find some structures composed of them behind the superficial phenomena. The decomposition is not always easy. And even when it is done well, we should not use the style of reductionists, but search for solutions or suggestions on the macroscopic level of phenomena re-constructing things by integration.

In the mid-90s we developed DEM (Discrete Element Model) to simulate macroscopic bed behaviors from laws governing microscopic phenomena. However, the computer capacity of that time was much smaller than now and it was difficult to conduct realistic scale computation. However, we conducted a variety of demonstration for beds of sticky particles to see agglomeration behavior, for beds of immersed tubes to see contact packet renewal over the tube surface, for beds of polymer particles in reaction to see distribution of temperature of individual particles, and for beds char combustion to see surface temperature of burning char which is important for NOx formation and reduction. With these small scale computation we could have tested innovative ideas in the macroscopic level, which is produced from freewheels thinking. Actually the real scale computation may not induce innovative ideas nor quick design as long as it takes much computational effort. Macroscopic models and correlations would be still valuable for quick actions as well as deep understanding of the phenomena of concern. Furthermore, with scaling laws we can conduct scaled down experiments as well as scaled-up computations for the integration.

2) Cluster and agglomerate size prediction

Prediction of mean cluster or agglomerate sizes is another challenge of macroscopic modeling. Here, I use the term a 'cluster' as a group of particles brought together by external (not particle-particle) effects and an 'agglomerate' as a group of particles held together by direct interparticle forces as proposed by myself and Professor R. Clift [3].

In the dilute region of CFBs and in the freeboard of bubbling beds we have clusters that are group of drifted particles. In bubbling beds of sticky particles we have agglomerates. In both cases the force fields changing dynamically are the key factors, one in a dilute suspension with turbulent gas motion and another in a dense bed with frequent compaction and sharing due to bubble passage. We have proposed models [4] and [5] for prediction of cluster and agglomerate sizes, respectively, but both models should be examined in more detail for validation.

4. Innovations

The key technological approaches for sustainability are low carbon transportation, zero emission buildings with highly insulated walls, biomass material utilization, solar heating, lighting and PV power generation, renewable energy utilization and process intensification, integration for low exergy losses. To reduce GHG emissions from transportation we need to avoid unnecessary carbon foot prints and induce local production and consumption. In terms of energy the scheme of decentralized local generation and consumption should be encouraged more to reduce emissions from centralized mineral energy conversion systems. Moreover, in many countries, their megalopolises consume much more energy than they can locally produce. In this respect encouraging people and industries move to areas rich in renewable energies is promising [6]. It is a counter action against the over-urbanization of the last 50 – 60 years. All of these indicate that decentralization and community empowerment should be the necessary trend in the coming decades for countries having centralized systems.

Destination of decentralization is also confirmed by thoughts on the global economy. So far the size of financial economy is said to be much larger than real economy. More money is invested for the sake of profit and there is a definite world trend of widening disparity and increase of poverty. Circulation of local investment in the local economy is one of the current issues to stabilize modern society. So, from economic viewpoint de-centralization of some extent and community economy empowerment are necessary.

Keeping these destinations in mind, the future practices of chemical and fluid-particle engineering should be newly defined. In the last part of my paper I would discuss their features-to-be.

References

1. M. Horio, Fluidization Science, its Development and Future, Particuology, 8, No.6, 514-524, 2010

2. M. Horio, 1. Overview of Fluidization Science and Fluidized Bed Technologies, 'Fluidized-bed Technologies for Near-zero Emission Combustion and Gasification,' ed. by F. Scala, Woodhead Publishing, 3-41, 2013

3. M. Horio, R. Clift, A Note on Terminology: 'Clusters and agglomerates', Editorial, Powder Technology, 70, pp.196, 1992

4. M. Horio, M. Ito, Prediction of Cluster Size in Circulating Fluidized Beds, J. Chem. Eng. Jp., 30, pp.691-697, 1997

5. Y. Iwadate, M. Horio, Prediction of Agglomerate Sizes in Bubbling Fluidized Beds of Group C Powders, Powder Technology, 100, pp.223-236, 1998

6. M. Horio et al., The Potential for Massive GHG Reduction by Mass Rural Remigration (The Renewable Energy Exodus): A Case Study for Japan, Applied Energy, 160, 623-632, 2015

474968 The Difference Between L-Valves and Loop Seals

Abstract:

People often do not understand the difference between operation of non-mechanical L-valves and loop seals, and this can lead to solids transport design failures. In their most common operating modes, an L-valve is used to control the flow rate of solids in a system, while the loop seal does not control the solids at all, but automatically moves solids automatically from a cyclone back into a fluidized bed.

Loop seals can operate with Geldart Group A, B or D solids if they are designed correctly. Only a little aeration is required on the upleg of the loop seal to maintain this part of the loop seal in a fluidized mode. The L-valve, however, is generally operated to control the solids flow rate around a system. Adding a little aeration to the L-valve at the correct location causes solids to flow through the Lâ€"valve. Adding more aeration causes a higher solids flow rate through the L-valve. While operating in this solids-control mode, the L-valve is practically limited to operating with solids greater than about 100 microns. Thus, only Geldart Group B and D solids can be used with an L-valve operating in the control mode. Group A solids do not work well with L-valves operating in the solids-control mode.

Another difference between the two devices is that the loop seal operates with an overflow, fluidized bed standpipe above it. The L-valve operating in a control mode cannot operate with this type of standpipe, but must operate only with an underflow, non-fluidized (packed bed) standpipe above it.

Why the L-valve operates differently than the loop seal will be explained by describing the differences in operation between the types of standpipes feeding the two non-mechanical devices.

468950 CFB Flow Regimes with Application to Fluidized Bed Combustion

Abstract:

Based on more than 25 years of fluidized bed (FB) research on mixing and fluid-dynamics with application to FBs in fuel conversion, we give an overview of the characteristics of gas and solids flow in fluidized beds for combustion and gasification. In particular, we discuss the flow and mixing characteristics of the four main regions identified in FB units for fuel conversion, with focus on CFB furnaces; the dense bed, the splash zone, the transport zone and the exit zone. We discuss the main characteristics of these zones and present recent results on how they relate to the net solids flux (G_s), i.e. the externally recirculated solids flux, which is a key parameter for CFB boilers and other dual bed systems, such as applied in chemical looping combustion for CO_2 capture and indirect biomass gasification. We draw our conclusions based on measurements and modeling of cold models as well as large commercial FB boilers and gasifiers.

CFB furnace characteristics

The furnace of fluidized beds for fuel conversion - combustion and gasification â€"typically have the following characteristics:

- Aspect ratio of the furnace (H_0/D_{eq}) of the order of or less than 10;
- Aspect ratio of the settled bed, i.e., the bed formed if the solids are not fluidized, $H_{b,settled}/D_{eq}$ of less than 1;
- Solids belonging to Group B in the Geldart classification; and
- Net solids flux that typically ranges from 0.5 to 20 kg/m^{2}s (CFB boilers).

Especially when it comes to circulating fluidized beds (CFB), experimental data on fluid dynamics and mixing reported in the literature were often carried out for other chemical engineering applications than fuel conversion, such as cracking, coating and polymerization. Such data may therefore not be used to draw conclusions of relevance for fuel conversion systems. In contrast to the characteristics listed above, the latter systems typically employ Geldart Group A particles and operate at net solids fluxes often far exceeding 20 kg/m^{2}s and for riser inventories which, if operating at low velocities, correspond to a tall bed operating in slugging regime - a regime which is not found in commercial-scale FBC boilers and gasifiers.

CFB furnace mixing and fluid-dynamics - an overview

In short, the fluid dynamics and mixing of the main fluid dynamic regions in a CFB furnace can be characterized as follows.

The dense bed: The dense bed is present provided there is a sufficiently high solids inventory and it is characterized by strong fluctuations in the gas flow due to the bubble passages. The dense bed can be maintained also when the superficial velocity exceeds the terminal velocity of a major part of (or all of) the bed solids (under CFB conditions). This is mainly due to that the primary gas passes through the dense bed in a heterogeneous way, partially in the form of large bubbles, which, at high velocities, create a shortcut of gas between the gas distributor and the region above the dense bed. Thus, this flow results in bubbles with oxidizing conditions and an emulsion phase under reducing conditions. Although the amplitude of the gas flow as induced by the bubble flow is drastically reduced above the dense bed, the gas flow fluctuations have important implications for the characteristics of the gas-solids flow in the freeboard.

The splash zone: Above the dense bed, there is a so-called splash zone that consists of solids, which is characterized by strong back mixing mainly due to ballistic movement of clustered solids thrown up by the bubble eruptions.

The transport zone: Under CFB conditions, the gas velocity is sufficiently high to cause a significant portion of the solids to be entrained up through the furnace. Thus, above the splash zone, a transport zone is formed which is characterized by that most of the back-mixing occurs at the furnace walls. The transport zone flow forms a core-annulus structure, with the up-flow mainly in the core region and with a continuous back-mixing at the furnace walls, thereby forming downward flowing wall-layers of solids. Such continuous back-mixing represents a solids size segregation up through the transport zone, i.e. the coarser the solids the higher the probability that they are separated at the furnace walls. Therefore, there is a continuous decrease in average solids size with height in the freeboard of a CFB furnace.

The exit zone: Near the inlet to the cyclone, an exit zone is formed which is governed by gas and solids flow properties and the design of the fluidized bed furnace. A certain portion of the solids which reach the level of the inlet to the cyclone are internally recirculated due to back-flow effects. Thus, whereas the above described flow characteristics of the bed, the splash and the transport zones seem to only depend on furnace (riser) pressure drop and fluidization velocity given a certain set of gas and solid properties, the externally recirculated solids â€" the net solids flux (G_s) â€" depend on exit geometry and the configuration and operation of the cyclone side of the CFB loop (cyclone dipleg and particle seal).

470434 Biomass Gasification in Fluidized Beds: Current Challenges and Advances

Abstract:

Fluidized bed technology applied to gasification of woody biomass can produce heat, power, liquid fuels, hydrogen and value-added chemicals with positive environmental impacts. Recent advances in this area include a culmination of fundamental and applied research leading to industrial development at pilot and demonstration scales. One of the hurdles affecting biomass utilization is the cost related to pre-treatment, biomass gasification, gas clean-up and downstream process. Novel technologies such as improved gasification conversion and destruction of tarts are required to improve the economics of the processes. Sorbent enhanced gasification advantageously combines CO2 capture with gasification. This talk highlights the state of the technology with experiences in tar cracking catalyst development, lime-enhanced gasification and community engagement through operation of two gasifiers at the University of British Columbia, Vancouver campus, while tracing the extraordinary contributions Prof. Horio has had in many aspects of the biomass utilization, fluidization and broader energy systems innovation.

454094 Flow Patterns in High Density Fluidized Beds Used As Solar Energy Carrier Systems

Abstract:

Dense gas-solid suspensions have the potential to be applied as heat transfer fluids (HTF) for energy collection and storage in concentrated solar power plants [1]. In the frame of the European CSP2 project [2], a solar receiver using a Dense Particle Suspension (DPS) upward flow in a bundle of tubes was studied. The interest of using as Heat Transfer Fluid (HTF) solid particles transported by an air flow is that the created two-phase medium, exhibits fluid-like flow properties while being able to withstand high temperatures up to the solid sintering temperature. Moreover, the DPS can be used as direct heat storage medium. In the design investigated here, the particles form a dense upward-flowing suspension. Both the density of the SiC particles suspension and their movement have a strong influence on the heat transfer, as demonstrated by Positron Emission Particle Tracking (PEPT) experiments [3].
The NEPTUNE_CFD computational code was used to simulate the DPS circulation both at ambient and high temperature. The numerical results are compared with those obtained on the cold mock-up and with PEPT experiments. The solid reflux evidenced by PEPT experiments was corroborated by simulations and during on-sun experiments [4].
Finally, the DPS flow is analysed highlighting the impact of the air velocity and the density variation along the tube height and the strong influence of the various operating parameters on the temperature distribution. The results of this study will be useful when planning for solar energy capture systems scale-up.

References:

[1] Gilles Flamant, Daniel Gauthier, Hadrien Benoit, Jean-Louis Sans, Roger Garcia, Benjamin Boissière, Renaud Ansart, Mehrdji Hemati, Dense suspension of solid particles as a new heat transfer fluid for concentrated solar thermal plants: on-sun proof of concept; Chemical Engineering Science 102, 567-576

[2] FP7 EC project CSP2. http://www.csp2-project.eu.
[3] Pablo Garcia-Trinanes, Jonathan Seville, Benjamin Boissiere, Renaud Ansart, Thomas Leadbeater, David Parker. Hydrodynamics and particle motion in upward flowing dense particle suspensions: Application in solar receivers. Chemical Engineering Science, (2016) 146:346-356.
[4] H. Benoit, I. PÃ©rez LÃ³pez, D. Gauthier, J.-L. Sans, G. Flamant. On-sun demonstration of a 750 C heat transfer fluid for concentrating solar systems: Dense particle suspension in tube. Solar Energy, (2015) 118:622â€"633

*Updated contact details: P.GarciaTrinanes@greenwich.ac.uk

469655 Time-Resolved X-Ray Tomography of Fluidized Beds

Abstract:

The dynamics of fluidized beds are challenging to study: the voidage distribution is changing quickly over time (i.e., a high temporal resolution is needed) and dense gas-solids flows are typically opaque to visible light. X-ray tomography can cope with these challenges. It is a non-intrusive technique that can be used for the reconstruction of the cross-sectional distribution of the different phases in fluidized beds. In Delft, we have an X-ray system consisting of three X-ray sources [1,2]. Originally, this system was equipped with 192 detectors, but we recently upgraded it to three vertical 30 x 30 cm detector plates with each 1524x1548 pixels, resulting in a resolution of about 1 mm.

We have applied our tomographic systems to a nanopowder fluidized bed. This is challenging, because of the much lower bulk density than for regular fluidized bed, as a result of the very high voidage [3]. We will show that the bulk density is a function of the height. Moreover, we will demonstrate that the history of the powder plays a strong role in nanopowders fluidization: stepwise increase of the gas velocity gives much poorer fluidization behaviour than starting with a high gas velocity, and stepwise decreasing it.

We also applied X-ray tomography to study the bubble hold up and its distribution in bubbling fluidized beds with and without vertical internals such as heat exchanger tubes. The goal is to understand the influence of vertical internals on the hold up. We show that with employed internals, the bubbles are more homogeneously distributed in the bed [4]. More, we have studied the influence of column diameter on bubble size and velocity in such a system [5].

[1] Van Ommen, J.R., Mudde, R.F., 2008, "Measuring the Gas-Solids Distribution in Fluidized Beds - A Review", International Journal of Chemical Reactor Engineering 6 (2008) R3.

[2] Mudde, R.F., Bubbles in a Fluidized Bed: A Fast X-Ray Scanner, AIChE Journal 57 (2011) 2684-2690.

[3] Quevedo, J.A., Omosebi, A., Pfeffer, R., "Fluidization enhancement of agglomerates of metal oxide nanopowders by microjets", AIChE Journal 56 (2010) 1456-1468.

[4] Maurer, S., Wagner, E. C., Schildhauer, T. J., van Ommen, J. R., Biollaz, S. M., & Mudde, R. F. (2015). X-ray measurements of bubble hold-up in fluidized beds with and without vertical internals. International Journal of Multiphase Flow 74 (2015) 118-124.

[5] Maurer, S., Gschwend, D., Wagner, E.C., Schildhauer, T.J., van Ommen, J.R., Biollaz, S.M., & Mudde, R.F., "Correlating bubble size and velocity distribution in bubbling fluidized bed based on X-ray tomography", Chemical Engineering Journal 298 (2016) 17-25.

461024 Integrated Exergy Recuperative Biomass Gasification and SOFC System for Hydrogen and Power Coproduction

Abstract:

Integrated Gasification Fuel Cell (IGFC) power generation system is being developed as high-efficiency power generation technology. However, the highest theoretical gross thermal efficiency of conventional IGFC is still below 52% because of large exergy destruction mainly due to partial oxidation gasification, air cooling of SOFC stacks, and the combustion in a gas turbine combustor. In order to obtain higher power generation efficiency, a novel hydrogen and power coproduction system based on the exergy recuperative biomass gasification integrated with solid oxide fuel cell was proposed. In the proposed system the exothermic heat from micro-tubular SOFC stacks immersed in the fluidized bed gasifier at a temperature range close to the SOFC operating temperature is recuperated and utilized for the endothermic heat of biomass gasification, preventing the exergy destruction due to the partial oxidation in gasifier and heat transfer for cooling SOFC. In addition, providing the recirculation system of unreacted hydrogen separated from SOFC anode exhaust, gas turbine can be eliminated. Since the exergy loss of SOFC is much smaller than that of gas turbine, the net power generation efficiency of the combination of gasification and SOFC power generation system can be considerably improved. Corresponding to this system, a novel triple-bed combined internal circulating fluidized bed gasifier, composed of a moving bed high-temperature pyrolyzer with the function of tar cracking, a bubbling bed char gasifier equipped with micro-tubular SOFC stacks, and a fluidized bed combustor of remained char, was proposed for 40kW class pilot plant. In the present study, the conceptual design of a integrated exergy recuperative biomass gasification and SOFC system for hydrogen and power coproduction by using a triple-bed combined internal circulated fluidized bed as a gsifier is described.

471392 Metal Oxide Reaction Engineering and Its Connection to Fluidization Systems for Energy Conversion Applications

Abstract:

Metal oxide reaction engineering has been widely applied in energy industries. A typical example of considerable current interest is chemical looping technology. Fossil fuel chemical looping applications were used with the steam-iron process for coal from the 1900s to the1940s and were demonstrated at a pilot scale with the carbon dioxide acceptor process in the 1960s and 1970s. There are presently no chemical looping processes using carbonaceous fuels in commercial operation. The enabling technology for chemical looping applications has been and will be the fluidization systems. A key factor that hampered the continued use of these earlier processes was the inadequacy of the reactivity and recyclability of the looping particles in the circulating fluidized bed operation. This factor led to unsustainable operation of the process. With CO2 emission control now of great concern, interest in chemical looping technology has resurfaced due to their unique ability to generate a sequestration-ready CO2 stream.

Chemical looping technology is a manifestation of the interplay among all the key elements of particle science and technology including particle synthesis, reactivity and mechanical properties, circulating fluidized bed stability and contact mechanics, gas-solid reaction engineering and particulates system engineering. This presentation will describe the fundamental and applied aspects of modern chemical looping technology that uses circulating fluidized bed processing fossil and other carbonaceous feedstock. Specifically, it will discuss the reaction chemistry, ionic diffusion mechanisms, metal oxide synthesis and thermodynamics, fluidized bed reactor design, and system engineering along with energy conversion efficiency and economics of the Coal-Direct Chemical Looping Process and Syngas Chemical Looping Process being developed at Ohio State University at a pilot level. Further, CO2 emission control using the chemical looping technology will be illustrated and compared with other CO2 capture methods. Potential for selective oxidation in the production of fuels and chemicals, as well as solar based chemical looping technology, will also be discussed.

451784 How Can Top People be Attracted ? Some Insight

Abstract:

Everyone would like to claim, that the organization they lead or are a part of, is a high performance organization. In reality, not too many are. Creating and nurturing a high performance organization starts with the hiring process. But hiring is only the first step. Once staffed, the organization has to walk the talk and deliver to the employees what was promised. If leaders do not deliver on their promises, good candidates are rapidly lost. In this session we would like to discuss key elements for sustainable success: forming organizations with happy, motivated, energized and contributing collaborators.

452868 Attracting and Nurturing Talent: Diversity, Inclusion, and Gender Empowerment in Industry

Abstract:

How do we change to make tomorrow's workplace welcoming to a diverse, inclusive, gender-empowered one? Attracting and nurturing talent is essential to the success of a team and to the ultimate success of a company. How do leaders in industry find, attract and develop talented contributors? Are we doing enough to attract smart, ambitious, energized, new employees? Is there a risk to the *status quo* if we explicitly make the effort to recruit, hire and develop a truly diverse workforce? We will pose these questions and discuss them.

461911 Challenges and Lessons Learned from Leading High Performing Cross Functional Teams

Abstract:

Cross-functional teams are significantly different from teams that are aligned on one functional level. Managing such teams with member highly accomplished in their own technical functions requires a careful leadership balance. This presentation will discuss what it takes to manage cross-functional. We will cover unique challenges and lessons learned from leading successful l cross functional teams in complex corporate environments.

How to Create and Maintain High Powered Teams: It Takes the Power of Influence

Kemi Sorinmade, The Growth Studio, Easton, MA

Introduction

When we think of high powered teams, teams which are high performing, we think of implementing management tactics, applying technical and functional knowledge, and following policies and procedures. Building a high performance team requires more than that. Leadership is usually the missing piece when it comes to achieving team success and effectiveness. When we think about leadership we think of success, but the way we lead has not resulted in the success we seek. This is because leadership is thought of to be as position, titles, strategy, and other autocratic methods. True and effective leadership involves influence, respect and driving results using non-traditional methods. In this paper, we will explore what true and effective leadership is, and expound on influence strategies that can be utilized by technical and engineering professionals to truly create synergy, improved performance and engagement at all levels. These principles will reveal for the reader, ideas and strategies that can be implemented to motivate and engage their teams for results.

The Role of Influence in Creating and Maintaining a High Powered Team

A true leadership culture leads to a successful team culture. Influence is at the core of leadership. Influence is the ability to achieve results and achieve change through others. Influence comes through respect and it is the highest form of leadership. A lot of teams don't succeed and perform well because they are run based on strategy, manipulation, force or position. These are all necessary but the missing piece for a successful team performance is influence. Teams comprise of people and people need to be led or influenced. Thus, growing and maintaining a great team takes good leadership. There are many facets to leadership and several qualities which define a good leader. But we will focus on the following traits to help us build high performance teams. Vision, Integrity, Empowerment, and Communication. Everyone has a part to play in influencing a high performance team. It is not the job of the positional leader alone.

What does a High Powered Team Look Like?

Teamwork is the center of great achievement. Teams provide, resources, energy, and more ideas, than one person can bring to the table. Each individual's potential is maximized and multiple perspectives from team members provide several solutions. Teams exist so that effort is multiplied and individual work is reduced. Teams exist to achieve a common goal or solve a problem that cannot be solved alone. Teams could exist in the form of a department, a project team, cross functional team or leadership team. Success for these teams could mean successful meetings, a successfully completed project, a problem solved, improved sales, satisfied customers, or highly engaged and motivated members. To achieve their goals, teams must perform well and drive results. A high powered team consists of individuals with a common goal and purpose, and who work together to achieve those goals. The following influencing strategies will help teams become more powerful and more cohesive.

Start with Purpose, Vision and Passion

Purpose

The foundation and success of a team lies in the purpose of the team. Why does the team exist in the first place? The purpose needs to be defined in terms of customers' needs. Never create a team because it's a requirement or a nice to have. Teams come about when there is a need. You may currently be leading a team or are part a team. Even for departmental teams which seem to exist naturally as part of an organization, a purpose needs to be established. Revisit why your team exists.

The vision

The vision is about where the team is going. Vision is such a common word and seems like common sense but a lot of teams still do not know where they are going. What is the end point and what are the goals? What do you want to be known as? What are you trying to accomplish? The answers to these questions need to be defined up front. As a leader of the team, you need to constantly share, in a clear fashion what these answers are i.e. your vision, goals and expectations. Without this your team members will lack initiative, drive, and motivation to follow through. These goals will be the drivers for everything you do: the daily activities, team meeting agendas, and they act as a filter for what you do or do not take on. These goals also help to determine who should or should not be on your team. When your goal is defined and constantly in front of you, everything falls into place.

When someone is not performing well as a team member, leaders need to ask themselves whether expectations have being shared with that team member. Meetings should be purposeful and agendas must always line up with the vision of the team. Meetings should always be an avenue for communication, recognition, problem solving, and creation of action items. If people don't leave the meeting with next steps or action items, then your meetings are not necessary.

With constant communication and reminders about goals and expectations, team members are not left in the dark, commitment increases and productivity increases.
Team members also have a part to play with regards to the goals and vision of the team. Team members should always ask for clarification in terms of their role and what is expected of them. They should also regularly ask for feedback, which leads to better performance.

When team members are united in purpose, there's no telling what they can achieve. This is why team leaders need to clarify and communicate vision and purpose constantly and allow time for everyone to buy-in.

Passion

Is everyone on the team passionate about their roles? Is the team leader or manager passionate about their role and team members? A lack of passion will lead to a lack of motivation and inspiration. The energy and passion a leader puts in is what they will get out of the team members. When a leader for instance is not passionate about a project in which they are the lead, the team members are not likely to be passionate. Another thing to consider is if the strengths of each member is aligned with the goals and work the team needs to perform? According to Steve Jobs, the only way to do great work is to love what you do. Working in one's strength and working on what you're passionate about directly correlate to potential. Play to the strength of each team member and ensure everyone that is on the team wants to be on the team.

Focus on Relationships

Relationships provide the glue that holds teams together. Improving relationships creates an open door for effective communication. Improved relationships create the synergy and connection required to get your team working successfully. As a team leader you have to build relationships with your team members and as a team member, you have to intentionally build relationships with your fellow team members. Relationships are more than just talking to one another. There is something known as the "know" "like" and "trust" factor. People will work with and do business with whom they know like and trust. To know, like and trust one another, relationships have to be built and maintained. Maintaining great relationships require a few rules of engagement. In other to engage with others you need to love and accept yourself first. When you are okay with who you are and you are confident with who you are, it is easy to reach out to others and give unto others. Building relationships is an unselfish act, it is about others and before you can make things about others you need to work on yourself first. Once as an individual, you have accomplished this you can then work on other things such as: 1) become more likeable and make it easy for people to work with you, 2) get rid of quirks or attitudes that repel others. 3) find out the names of people and remember those names, 4) focus on others' interests, 5) put others ahead of you 6) request help as often as you can 7) find common ground and connect with those you think are different from you, and 8) always show appreciation.

The success of a manager, leader of a team, or a team member depends more on relationships than on technical or functional skill.

Trust, the Foundation of High Powered Teams

People will do business and work with whom they know like and trust as mentioned earlier. Trust is a big factor that determines whether people will want to work together. Integrity, character, and credibility make trust possible.

When everyone on a team works on their individual character and build their credibility, the sky is the limit for the success of your team. Integrity commits itself to character over personal gain, to people over things, to service over power, to principle over convenience, and to the long view over the immediate. Integrity is about telling the truth, keeping promises to ourselves and others, and being accountable. As a result, each individual is known for their consistency and discipline which in turn leads to trust. People begin to trust each other. Personal integrity also creates an environment in which people can easily ask questions, share ideas and are allowed to take risks and even fail.

People do What People See

Become and model what you want to see. People will do what people see. Instead of pointing fingers and placing blame, team members should become good examples and model what they want to see in others. As a team leader or team member, your job is to be an example. People respond based on what they see and not on what you say. If something is not working right instead of complaining, do what you can to model the right solution to the problem. Integrity helps individuals lay the right foundation for building credibility. When people keep promises and remain disciplined enough to complete tasks it results in a good track record for that team member which in turn creates trust and dependability. Team members need to depend on one another to get work done.

Integrity is the responsibility of each team member

When each team member takes responsibility for maintaining integrity and character, it increases the speed by which trust is present among team members. Here are some ways team members can build integrity and character

- ✓ Adhere to ethical and moral standards
- ✓ No lying, no stealing, no cheating
- ✓ Commit yourself to what is right over personal gain
- ✓ Don't make commitments you can't keep
- ✓ Keep commitments you have made
- ✓ Be respectful
- ✓ Be authentic. Don't play politics, don't role play
- ✓ Be disciplined
- ✓ Be truthful even when it hurts
- ✓ Complete tasks and assignments

Core Values

Your core values are the guiding principles for your team. These stem from purpose and they help to lay the foundation for a good team culture. What will your team be known for? What guides your decisions? A team with strong core values which are understood, shared and embraced by each team member is destined for success. A team's commitment to integrity and character paves the way for the development of these core values. Great examples of core values are: integrity, commitment, service, communication, responsibility, etc. Of course each team has to develop their own set of core values and these should guide every decision, every activity, and action of each team member. Decisions are made faster and everyone stays on the same page.

Embrace Empowerment

To empower means to give power to others. The main ingredient for empowering others is a high belief in people. If you believe in people they will believe in themselves and perform to the best of their abilities. Each team member has the responsibility to see others in terms of their potential not where they are right now.

As the team leader, ask for input and feedback. This requires humility, character and self - acceptance. Have people take turns in leading meetings or assign different leadership roles to everyone on the team. As an individual team member, always seek to help others and ask for help in return. If you see any team member struggling, provide them with tools and resources to help them. Give part of your job to others. This creates a sense of importance for each team member and no one feels left out. It leads to creativity, innovation and multiplication of effort.

Drive Results

The success of a team is characterized by its ability to achieve results. To achieve results, teams have to take action, and action leads to results. The problem is that teams are usually working but are working on the wrong things. Strategies are developed, plans are made, but they are not followed. To drive results, a high performing team must know how to prioritize. This goes back to the vision, purpose and core values. Everything a team does, action items and plans should derive from these. To develop the right action items, three questions need to be answered. Which activities line up with our goals? Which of these activities yield the best results? Who is the best person to carry out each of these activities? The team should be spending eighty percent of their time and resources on these defined activities. Meetings should also follow these same rules.

Team Building and Performance Checklist

- ☐ The purpose of the team has been established
- ☐ Expectations are clear for each person on the team
- ☐ The team leader models what is expected.
- ☐ Communication of vision, expectations, feedback are clear, regular, and done in a timely manner
- ☐ The team leader appreciates team members and team members appreciate one another
- ☐ Team members freely help each other
- ☐ Team leader accepts blame and gives credit
- ☐ Tools and resources needed for each member to function are provided
- ☐ Each member of the team is playing to their strengths
- ☐ An evaluation is done periodically to make sure that everyone that is on the team wants to be on the team
- ☐ Team members are held accountable
- ☐ The team environment is one where people are free to trust
- ☐ Team members are playing to their strengths and released from their weaknesses
- ☐ Personal relationships are maintained and improved upon
- ☐ Members connect before getting into work matters

References

1. Maxwell, John C., "The 21 Irrefutable Laws of Leadership," Thomas Nelson, 2007.
2. Maxwell, John C., "The Complete 101 Collection. What Every Leader Should Know," Thomas Nelson 2010.
3. Maxwell, John C., "Becoming a Person of Influence", Thomas Nelson 2007.
4. Lencioni, Patrick, "The Five Dysfunctions of a Team" Jossey-Bass, 2002.

473344 Developing High Performance Teams in Emerging Non-Traditional Entreprenurial Operations

Abstract:

Entrepreneurial venture participants are either superstars or oddballs. Both traits are needed. The effort in building a high performing team must focus on finding the mix of innovative off the wall creativity, with the ability to drive fast-paced targets and get things done. Decisions must be made on the fly without all the information, but the aspects of safety and ethics can never be compromised. Innovation is unleashed by having mundane tasks taken out of the equation.

An entrepreneurial team is built on excellence, confidence and passion. Changing the world comes with with chaos. Relationship is built on the abnormal. A high performing team is built on trust. Non-traditional members of a venture may not be viewed as the traditional high performers, but they are high performers by what they deliver. The same traits that make it great also tends to blow it apart.

The emerging engineering and operations focus on non-traditional chemical engineering feedstocks has brought many unbeknownst to them non-traditional resources into traditional chemical engineering operations.

The challenge is harnessing this new passion. There is nothing more fulfilling, exciting or satisfying.

472903 Evaluating Electrocatalysts for Solar Fuels: Experimental Methods, Performance Metrics, and Catalyst Design Strategies

Abstract:

Electrocatalysis is a key aspect that enables a number of technologies for the renewable, sustainable production of fuels and chemicals. Examples include photovoltaic coupled water electrolysis (PV-electrolysis) and solar photoelectrochemical (PEC) water-splitting, where catalysts are needed for the hydrogen evolution reaction (HER) and the oxygen evolution reaction (OER). Related technologies involve catalysts for CO2 electro-reduction to produce carbon-based fuels and chemials. This presentation will cover several aspects regarding electrocatalysis for these reactions, in particular experimental methods, figures of merit for comparing catalysts, and resulting design strategies for future catalyst development.

474117 Electrochemical Pathways for Sustainable Manufacturing

Abstract:

The chemical industry represents a key component of economic development and prosperity worldwide. However, they confront technology challenges â€" e.g., reliability of energy supply, lack of energy efficient/transformative manufacturing technologies, waste reduction, and water conservation- that hinder and jeopardize their growth and that could affect their competiveness worldwide. Current chemical industrial production methods have approached their practical performance limits, therefore, new transformational, disruptive, and enabling technologies are needed that will provide solutions to the chemical industry beyond incremental manufacturing improvements. Furthermore, a large and growing number of manufacturers are realizing substantial financial and environmental benefits from sustainable business practices. Sustainable manufacturing is the creation of manufactured products through economically-sound processes that minimize negative environmental impacts while conserving energy and natural resources.

Electrochemistry provides the potential to develop transformational technologies to make more efficient traditional chemical and biological processes. This talk will discuss examples of advanced applications of electrochemical science and technology for the sustainable production of ammonia (chemical), graphene (advanced material), and water remediation.

474722 Challenges and Current Development of Sulfur Cathode in Lithium-Sulfur Battery

Abstract:

Rechargeable lithium-sulfur (Li-S) batteries have exceptional theoretical capacity; however, the practical applications are still elusive to date due to many critical challenges. This presentation summarizes the technical issues of the sulfur cathode and the state of the art strategies in recent years to address these issues from the aspects of lithium polysulfides sequestration, new concepts of sulfur cathode materials, functionality of binders, and the roles of the electrolytes. With regard to conventional sulfur cathodes, liquid-phase reactions involving polysulfides are inevitable. Therefore, the cycle stability is essentially dependent on the affinity of the cathode with lithium polysulfides. Cathode components, including the sulfur host, additive, and binder, should all be active to attract lithium polysulfides. Because polysulfide anions are the electrochemically active species, the sequestration functionality of the cathode components should be specifically designed to attract polysulfide anions. Metal oxides and heteroatoms show promising properties in attracting polysulfides, but the electrochemical â€œinactiveâ€ mass associated with their use needs to be minimized. Polycation binders also seem effective in their ability to attract polysulfide anions due to the electrostatic attraction, but systematic investigation on binders is lacking. In addition to the polysulfide affinity, other properties of polymer binders, including mechanical strength (molecular weight), solubility, and long-term stability in specific electrolytes, are all important to the optimization of the binder system. Moreover, the electrolyte/sulfur weight ratio needs to be watched carefully since almost all studies in open literature used a ratio too high to ever achieve full cell capacity to surpass state of the art Li-ion batteries. It is clear that the development of practical high capacity Li-S batteries still has a long winding road ahead to travel.

457098 Mathematical Modeling of Lithium Ion Cells and Batteries

Abstract:

A physics-based models of lithium ion cells can be used to make predictions of the voltage of the cell as a function of time for charge and discharge given the design parameters and operating conditions of the cell. Such a model enables cell designers to determine the effect of changing design parameters on cell performance before building the cell for a particular application. These models can also be used to design a thermal management system to ensure that the heat generated in the cell is removed before causing overheating and thermal runaway of the cell. These physic-based models of a lithium ion cell can be extended to include multiple spatial dimensions to predict the temperature distribution in a lithium-ion cell for a given set of conditions. A review of the development of physics-based lithium ion cell models will be presented, with particular attention given to predicting temperature distribution in batteries and the effect of volume expansion on battery performance.

475206 Understanding and Overcoming the Body's Biological Barriers for Drug Delivery

Abstract:

Effective delivery of drugs is a major problem in today's healthcare. At a fundamental level, the challenge of drug delivery reflects the fact that the drug distribution in the body is limited by body's natural metabolic processes and transport barriers. These biological barriers, while serving an important purpose of regulating body's metabolic functions, limit the drug dose that ultimately reaches the target site. Accordingly, many drugs fail to reach their full therapeutic potential. Our research aims at developing a fundamental understanding of body's key biological barriers such as skin and intestinal epithelium, and utilizing this understanding to develop novel means to negotiate these barriers to deliver drugs.

Human skin is one of the most sophisticated and well-engineered biological barriers in the human body. In principle, it offers an ideal interface to administer drugs into the body through the use of a transdermal patch. However, its formidable barrier properties limit the drug dose that can enter the body. Our research has led to the understanding of how skin's structure and transport properties can be modulated using external stimuli such as ultrasonic waves, fluid microjets and amphiphiles, and how these stimuli can be controlled to enable transdermal delivery of drugs that were once thought undeliverable. We have also extended the lessons learned from our skin exploration to understand and negotiate other biological barriers in the body, in particular intestinal epithelium, reticuloendothelial system and tumors. Using transdermal-inspired approaches, we have advanced ways to understand and enhance drug delivery across the intestinal epithelium via oral route using patches. We have also designed patches that can be delivered within the body using circulatory cells such as red blood cells and macrophages by navigating through the internal biological barriers of the liver, lung, brain and tumors. I will present an overview of the lessons learned from our exploration of these biological barriers.

475213 Advances in 3D Manufacturing

Abstract:

3D printing has struggled to deliver on its promise to transform manufacturing. Prints take forever, parts are mechanically weak, and material choices are far too limited. In this talk I will update a breakthrough technology called Continuous Liquid Interface Production (CLIP) that grows parts instead of printing them layer by layer. CLIP allows businesses to produce commercial quality parts at game-changing speeds, creating a clear path to 3D manufacturing.

475214 Additive Manufacturing and Architected Materials

Abstract:

Material properties are governed by the chemical composition and spatial arrangement of constituent elements at multiple length-scales. This fundamentally limits material properties with respect to each other creating trade-offs when selecting materials for specific applications. For example, strength and density are inherently linked so that, in general, the more dense the material, the stronger it is in bulk form. We are combining advanced microstructural design, using flexure and screw theory as well as topology optimization, with advanced additive micro- and nanomanufacturing techniques to create new material systems with previously unachievable property combinations â€“ mechanical metamaterials. The performance of these materials is fundamentally controlled by geometry at multiple length-scales rather than chemical composition. We have demonstrated designer properties of these mechanical metamaterials in polymers, metals, ceramics and combinations thereof. Properties include ultra-stiff lightweight materials, negative stiffness, and negative thermal expansion. Our additive micro-manufacturing techniques include Projection Microstereolithography (PïSL), Direct Ink Writing (DIW), and Electrophoretic Deposition (EPD). These tools are capable of generating the designed structures which are highly three-dimensional micro- and nano-scale architectures with multiple constituent materials in the same structure.

475215 Additive Manufacturing/3D Printing from a Polymer Materials Manufacturer's Perspective

Abstract:

Eastman Chemical Company is a large, multinational polymer materials manufacturer currently exploring new materials for additive manufacturing (AM) in the fused filament fabrication (FFF)/fused deposition modeling (FDM) and laser and infrared-based powder sintering areas. When examining 3D printing, many similarities to the industrial development/build-out of injection molding can be found. Both techniques started with rudimentary equipment and, especially, limited materials that hindered true large-scale manufacturing market activation. While it took the dire manufacturing needs of a world war to activate new injection molding systems and polymer materials, modern societal and market drivers such as increased productivity, local production, just-in-time inventory, and individualization are driving additive manufacturing forward. After decades of relatively closed hardware and materials options, open materials approaches by major new 3D printer makers coupled with powerful design-oriented software are finally delivering the tools to rapidly boost end users past the so-called "hype phase" into true manufacturing-scale sectors. I will talk about Eastman's efforts to commercialize our materials into the 3D printing space, the upcoming resurgence of FFF printing, why massive parallel AM printing might eventually meet (or even beat) injection molding process speeds, the need to take AM truly to a continuous process, and how we see previously fragmented market areas beginning to focus on tooling, molding, and more personally, unmet needs in sectors such as prosthetics. While 3D printers may not end up in every home, there are solid, new businesses with additive manufacturing as their core just around the corner!

476796 HP's Jet Fusion 3D Printing Technology: Enabling the Next Industrial Revolution

Abstract:

In 2014 Hewlett-Packard announced the development and commercialization of an innovative 3D printing technology that promised to set new standards for performance, quality, reliability and low TCO. HP's Multi Jet Fusion™ (MJF) technology achieves its breakthrough performance by leveraging the company's 30+ year history of innovation and market leadership in imaging and digital printing. This presentation will provide an introduction to a ***new-to-the-world*** digital fabrication technology that makes it possible to design and print three-dimensional objects that possess both precise geometric and functional characteristics. The MJF technology will radically change the way engineers and designers prototype and produce functional parts and the blending of HP's MJF 3D printing technology with ***digital materials design*** creates a new fabrication paradigm - a paradigm that enables innovation in both form and function.

461114 Solid-in-Oil (S/O) Nanodispersions for Transdermal Cancer Immunotherapy

Abstract:

Transcutaneous immunization is a novel, non-invasive alternative to conventional immunization by injection. Skin immunocompetence comprised of abundant antigen-presenting cells in the epidermal and dermal layers of the skin can provide an effective tool for transcutaneous immunization, whereas the outermost, hydrophobic layer of skin, the stratum corneum, hinders penetration of antigens into the skin. To realize effective transcutaneous delivery of antigens, we have developed a solid-in-oil (S/O) technique that produces an oil dispersion of hydrophilic biomolecules.

In this study, we applied this technique to a cancer antigen and aimed to develop an effective transcutaneous cancer vaccine. We chose the melanoma, a kind of the skin cancer, to evaluate the cancer immunity. Because a melanoma antigen, TRP-2 peptide, has low solubility in water, we added lysine residues to the N-terminus of the peptide and the newly developed peptide antigen, K-TRP-2, was used as a melanoma antigen. In addition, we tried the use of R-848 as hydrophobic adjuvant to induce stronger immunity. The *in vivo* experiment revealed the effective prevention of tumor and the metastasis was achieved by S/O formulation containing R-848.

473822 Micromotors Generating Self-Propelled Regular Motion By Organic Fuels

Abstract:

Most catalytic micro/nanomotors that have been developed so far use hydrogen peroxide as fuel, while some use hydrazine. These fuels are difficult to apply because they can cause skin irritation, and often form and store disruptive bubbles. In this paper, we demonstrate a novel catalytic Pt micromotor that does not produce bubbles, and is driven by the oxidation of stable, nontoxic primary alcohols and aldehydes with dissolved oxygen. This use of organic oxidation mirrors living systems, and lends this new motor essentially the same characteristics, including decreased motility in low oxygen environments and the direct isothermal conversion of chemical energy into mechanical energy. Interestingly, the motility direction is reversed by replacing the reducing fuels with hydrogen peroxide. Therefore, these micromotors not only provide a novel system in nanotechnology, but also help in further revealing the underlining mechanisms of motility of living organisms.

473467 Pattern Formation of Interfacially Confined Periodically Sequenced Polypeptides

Abstract:

Periodically sequenced peptides can be confined to interfaces and assembled into patterns that present chemical functionalities with exceptional spatial precision. These rationally designed peptides and polypeptides are rapidly becoming useful components in nanostructured materials for applications ranging from drug delivery to energy storage. This presentation will examine several fundamental aspects of self-assembly and pattern formation of well-defined sheet forming peptides confined at interfaces. Our approach involves three steps. (1) We design and synthesize simple periodic peptide sequences, yielding surface-active Î^2-strands that self-organize into aggregates to form patterns as a function of the peptide sequence. Rational peptide design allows us to systematically explore the role of hydrophobicity, electrostatics and molecular size on materials properties. (2) We use a set of interfacial characterization tools to examine in intermolecular assembly and supramolecular mechanics of the self-assembled structures. (3) We apply two-dimensional equations of state that define both the phase behavior and the critical surface concentrations of nascent aggregates at the interface. Subsequently, we can apply these parameters to predict the dimensions of pattern formation and to determine the potential of the peptide assemblies as biomimetic materials.

474938 Self-Assembly Under Conditions Mimicking Deep-Sea Hydrothermal Vents

Abstract:

Deep-sea hydrothermal vents are hot spring occurring in the bottom of the ocean. Hot and compressed water, sometimes in the supercritical state (T_c = 374 °C, P_c = 22.1 MPa), gushes out from the sea floor and rapidly cooled to near-freezing temperature by surrounding cold water. In this communication, self-assembly under such unique conditions and its possible implication for the origin of life will be discussed.

Preparation and Properties of Niosomes Prepared with Polyglycerol Fatty Acid Esters Using the Supercritical Carbon Dioxide Reverse Phase Evaporation Method

Shunsuke Yamaguchi [1], Kenichi Sakai [2),3)] and Hideki Sakai [2),3)]

1). NIKKOL GROUP Cosmos Technical Center Co., Ltd., 3-24-3 Hasune, Itabashiku, Tokyo 174-0046, Japan

2). Department of Pure and Applied Chemistry, Faculty of Science and Technology, Tokyo University of Science, 2641 Yamazaki, Noda, Chiba 278-8510, Japan.

3). Research Institute for Science and Technology, Tokyo University of Science, 2641 Yamazaki, Noda, Chiba 278-8510, Japan.

Keywords: nonionic surfactant, polyglycerol fatty acid ester, supercritical carbon dioxide reverse phase evaporation method, niosome, liposome

Abstract

We have previously reported a new preparation method of liposomes and niosomes (nonionic vesicles) using supercritical carbon dioxide (scCO_2) as a solvent (scRPE method). In this study, niosomes were prepared by polyglycerol fatty acid ester (PG ester)-type nonionic surfactants, which are made from naturally-derived materials and have a high safety for a human body and environment, using the scRPE method with ethanol as co-solvent. Decaglycerol distearate (DG2S) and decaglycerol diisostearate (DG2IS) successfully formed niosomes using the scRPE method. On the other hands, decaglycerol monostearate (DG1S) with a high hydrophilic-lipophilic balance (HLB) value gave a spherical micellar solution, and decaglycerol tristearate (DG3S) with a low HLB value gave a gel-like solution. Niosomes of DG2IS had higher trapping efficiency and dispersion stability than DG2S niosomes because the membrane fluidity of DG2IS niosomes was higher than that of DG2S niosomes. The niosomes obtained in the present study would be a useful candidate for applying to cosmetics and drugs because they are formed by nonionic surfactants derived from natural sources and are prepared using the scRPE method without harmful organic solvents.

1. Introduction

Liposomes have phospholipid bilayers with an aqueous phase as their core[1], and have been used as a drug carrier in drug delivery systems, cosmetics, a biomembrane model, a microfield for chemical reactions, and so on[2-8]. One of the most peculiar properties for liposomes lies in retaining water-soluble substances in the inner aqueous phase and oil-soluble substances in the bilayers[9, 10].

Although liposomes are expected to be practically used for wide industrial fields, there are several problems in terms of physical and chemical stability. One of the most useful approaches to improve this problem is a usage of nonionic surfactants in place of phospholipids to form vesicles. They are suitable for industry applications compared with liposomes because of their high physicochemical stability and low cost performance. Furthermore, there are a large number of nonionic surfactants available for the design of nonionic vesicles, niosomes, because of many combinations of hydrophilic groups and lipophilic groups. For instance, there have been many reports to prepare niosomes using polyoxyethylene alkyl ether, sucrose fatty acid ester, polyglycerol fatty acid ester, sorbitan fatty acid ester and polyoxyethylene sorbitan fatty acid ester[11, 12]. However, a variety of preparation methods to obtain liposomes or niosomes require for the addition of harmful organic solvents like chloroform or acetone in their preparation process for dissolving phospholipids or nonionic surfactants[1, 14, 15]. When vesicles are applied to drug carriers and cosmetics, use of organic solvents must be avoided as much as possible. Recently, we have reported about a new environmentally-friendly method, the supercritical carbon dioxide reverse phase evaporation method (scRPE method), for the preparation of liposomes or niosomes in a single step using supercritical CO_2 ($scCO_2$)[16, 17]. However, the niosome formation with the scRPE method has been confirmed only for nonionic surfactants with polyoxyethylene as their hydrophilic groups[18,19]. In pharmaceutical, cosmetic and food fields, consumers have tendency to demand to use products made from more safe and plant-derived materials. Thus, the usage of materials containing petroleum-derived materials like polyoxyethylene-type nonionic surfactants is a tendency to be avoided. However, there were a few reports about niosomes prepared with non-polyoxyethylene-type nonionic surfactants.

In this study, the scRPE method is applied to prepare niosomes with polyglycerol fatty acid ester (PG ester)-type nonionic surfactants. PG ester-type nonionic surfactants are widely used as food additives due to their higher safety. We report in this paper the effects of molecular structure of nonionic surfactants on the niosome formation and their physicochemical properties in terms of trapping efficiency measurements, cryogenic transmission electron microscopy (cryo-TEM), fluorescence anisotropy (FA), differential scanning calorimeter (DSC) and dynamic light scattering (DLS).

2. Experimental

2.1. Materials

Decaglycerol monostearate (DG1S), decaglycerol distearate (DG2S), decaglycerol diisostearate (DG2IS) and decaglycerol tristearate (DG3S) listed in Table 1 provided by Nikko Chemicals (Tokyo, Japan) were used as PG ester-type nonionic surfactants. These materials were derived from natural sources and were used without further purification.

Table 1. Polyglycerol fatty acid ester-type nonionic surfactants used in this study

Chemical structure of nonioinic surfactants	HLB*
Polyglycerol fatty acid ester 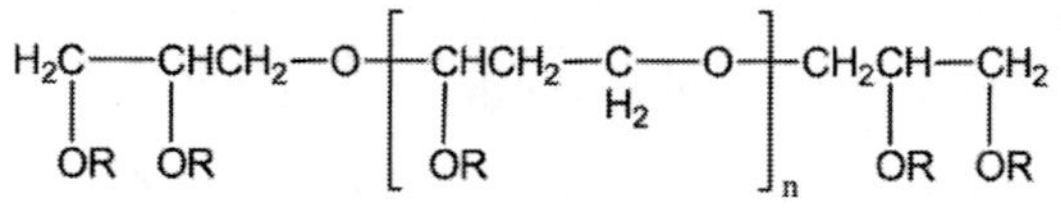 n = ca.10 R = H or stearyl, isostearyl (esterification degree = 1 to 3)	
Decaglycerol monostearate (DG1S)	12.0
Decaglycerol distearate (DG2S)	9.5
Decaglycerol tristearate (DG3S)	7.5
Decaglycerol diisostearate (DG2IS)	10.0

* HLB values were given by suppliers.

2.2. Methods

2.2.1 scRPE Method

Fig. 1 shows a schematic illustration of the experimental apparatus for the scRPE method. After a nonionic surfactant and ethanol at desired amounts were sealed in the cell, CO_2 was introduced into the cell. The cell temperature was then raised to 60°C, while the pressure was kept at desired CO_2 pressure. After a few minutes for equilibration, aqueous solution of 0.2 mol/L glucose, a water-soluble model drug substance, was slowly (0.1 mL/min) introduced using a HPLC pump until total volume of 5 mL and the surfactant concentration of 10 mmol/L. The pressure was then reduced to release CO_2, thereby giving suspensions. The interior of the cell was stirred with a magnetic stirring tip during the preparation process.

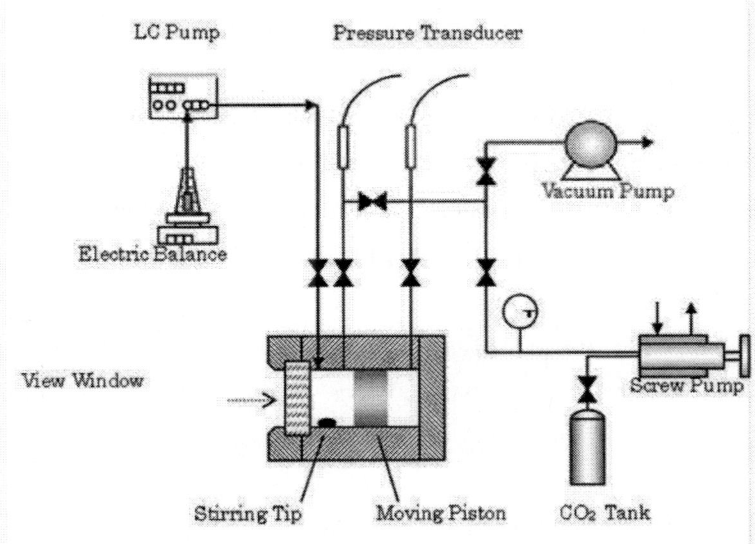

Figure 1. Apparatus for preparation of niosomes with the scRPE method

2.2.2 Evaluation of niosome suspensions

For determination of trapping efficiency of glucose as a water-soluble model drug entrapped in niosomes, the glucose-loaded niosome suspension was dialyzed against water using a cellophane tube (Viscase Scales Co.) to remove unentrapped glucose. The niosomes inside the tube were destroyed by the addition of ethanol. The amount of glucose in the solution was determined by the mutarotase GOD method[20, 21)] using a spectrophotometer (UV-260, Shimadzu Co.).

The hydrodynamic diameter of niosomes was estimated using a NICOMP 380ZLS particle size analyzer equipped with a 5 mW He-Ne laser at a constant detector angle of 90°. The obtained scattering data were fitted using a number-weighted NICOMP-mode analysis to estimate the diffusion coefficient of the surfactant assemblies in the solution. The hydrodynamic diameter was obtained from the diffusion coefficient using the Stokes-Einstein equation.

The formation of niosomes was confirmed by cryogenic transmission electron microscopy (cryo-TEM). A small amount (3–5 μL) of sample solution was placed on the surface of a TEM copper grid covered by a holey carbon film, which was held by a pair of self-locking tweezers mounted on a spring-loaded shaft with a cryo-preparation system (LEICA EM CPC, LEICA microsystems). The sample drop was blotted with a filter paper to form a thin liquid film on the grid (<300 nm), and immediately plunged into liquid ethane cooled by liquid nitrogen (<–170°C). The grid was transferred onto the tip of a cryospecimen holder (CT-3500, Oxford Instruments) under liquid nitrogen. Specimens are kept at temperatures less than –170°C and imaged in a transmission electron microscope (H-7650, Hitachi Science Systems, Ltd.) at an accelerating voltage of 120 kV under low electron dose.

DSC measurements were performed with a differential scanning calorimeter (DSC-60, Shimadzu Co.) at a heating rate of 5 K/min. A stainless steel pan was used as the sample vessel. The reference was distilled water with the same weight as that of the sample.

Examination of the microviscosity of bilayers was made through fluorescence polarization measurements using 1,6-diphenyl-1,3,5-hexatrience (DPH) as a fluorescence probe. DPH was dissolved in tetrahydrofuran, and the resultant solution was added to each of the niosome suspensions to make the molar ratio of DPH to surfactant 1:100. The DPH containing solutions were then incubated at 37°C for 2 h to solubilize the probe and the fluorescence intensity was measured with a fluorescence spectrophotometer (RF-5000, Shimadzu Co.). The excitation and emission wavelengths were 350 and 450nm, respectively. Fluorescence polarization (P) was calculated using the following equation.

$$P = (I_p - I_v)/(I_p + I_v)$$

where I_p and I_v are the fluorescence intensities of the emitted light polarized parallel and vertical to the exciting light, respectively.

3. Results and Discussion

3.1. Solubility of nonionic surfactants in $scCO_2$ and its dependence on ethanol addition

From our previous study[19], it is necessary that nonionic surfactants are completely dissolved in $scCO_2$ and that their hydrophilic and lipophilic balance (HLB) ranges from 9.5 to 9.9 in order to form niosomes using the scRPE method. All nonionic surfactants used in the present study showed a poor solubility in $scCO_2$ at 60°C without co-solvents. Furthermore, in the case of without ethanol, these suspensions did not have trapping efficiency. The addition of ethanol as co-solvent[17] is thus required for the formation of niosomes using these nonionic surfactants. The addition of ethanol is allowed in the fields of cosmetics and pharmaceutical industries if the amount added is proper. Fig. 2 shows the solubility of the nonionic surfactants in $scCO_2$ at 60°C as a function of the amount of ethanol. In this figure, one phase region means complete dissolution of surfactants in $scCO_2$, while two phases region means that a part of the surfactants added cannot dissolve in $scCO_2$. DG1S did not completely dissolve in $scCO_2$ below 200 bar (the upper limit of CO_2 pressure in the present equipment) even if ethanol was added in the surfactant solution. On the other hands, DG2S was soluble in $scCO_2$ at the pressure larger than 130 bar when the concentrations of ethanol were more than 7.5 wt.%. In the case of DG2IS, the complete dissolution was confirmed at ethanol concentrations larger than 2.5 wt.% and at the CO_2 pressure larger than 120 bar. For DG3S system, the dissolution was confirmed at ethanol concentrations larger than 5.0 wt.% and at the CO_2 pressure larger than 105 bar. This result showed that the solubility in $scCO_2$ is increased with increasing the esterification degree of PG ester-type surfactants, because the solubility of $scCO_2$ was close to hexane and $scCO_2$ was easy to dissolve lipophilic surfactants. In the following experiments, on the basis of this result, niosomes were prepared at 60°C, at 7.5 wt.% ethanol concentration, and at 130 bar of CO_2 pressure where DG3S, DG2S and DG2IS were soluble in $scCO_2$.

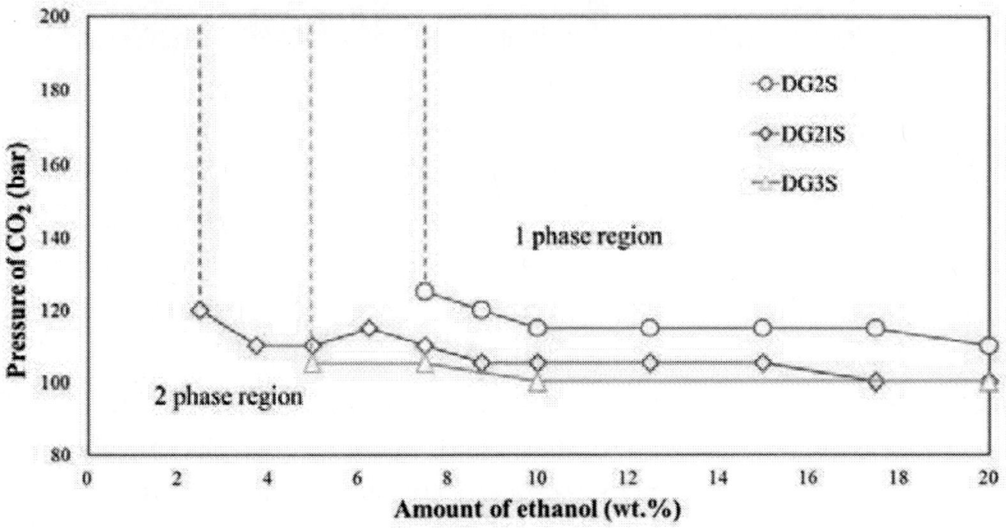

Figure 2. Phase diagram of polyglycerol fatty acid ester-type nonionic surfactants (DG2S, DG2IS and DG3S) in $scCO_2$ as a function of the amount of ethanol at 60°C. DG1S was insoluble in $scCO_2$ at every pressure of CO_2 and ethanol concentration investigated in this study.

3.2. Preparation of niosomes using PG ester-type nonionic surfactants

Preparation of niosomes using PG ester-type nonionic surfactants were conducted by the scRPE method at the proper condition described in 3.1. Table 2 shows the results of visual observation, average particle size and trapping efficiency for the PG ester-type nonionic surfactant suspensions just after preparation. Turbid suspensions were obtained using DG2S and DG2IS, which suggests the formation of niosomes. On the other hands, usage of DG1S and DG3S gave transparent and gel-like solutions, respectively. The average particle size for DG1S was less than 10 nm, which indicated the spherical micelle formation. The aggregates of DG2S and DG2IS had particle sizes of 340 nm and 220 nm, respectively. In addition, the trapping efficiency values for DG2S and

DG2IS were 5 and 20%, respectively, while the DG1S suspension had almost zero % as trapping efficiency.

Table 2. Appearance, average particle sizes, and trapping efficiency for PG ester-type nonionic surfactant suspensions using the scRPE method

	DG1S	DG2S	DG3S	DG2IS
Visual observation			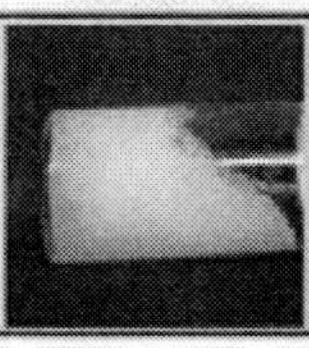	
	Translucent liquid	Turbid liquid	Gel	Turbid liquid
Particle size (nm)	9	340	N.A.	220
Trapping efficiency (%)	0	5	N.A.	20

Figure 3 shows cryo-TEM micrographs for molecular assemblies formed by DG2S and DG2IS. Hollow structures observed in the cryo-TEM micrographs strongly suggest the formation of niosomes for DG2S and DG2IS systems. On the other hands, although DG3S was dissolved in $scCO_2$ by the addition of ethanol as co-solvent, the formation of niosomes was not confirmed and gel-like solution was obtained. Our previous report[19] shows that proper HLB values of nonionic surfactants for the niosome formation range from 9.5 to 9.9. For DG2S and DG2IS, whose HLB values are in or similar to the proper HLB range, niosomes were formed using the scRPE method. Because the HLB value (7.5) of DG3S is smaller than the lower limit of this proper HLB range, lamellar structures were not stably formed for DG3S. In the case of DG1S, the HLB value (12.0) is higher than the upper limit of the proper HLB value for niosome preparation. These findings obtained here showed that high solubility of the PG ester-type nonionic surfactants in $scCO_2$ by the addition of ethanol, as well as usage of nonionic surfactants with the proper HLB value, are required for the formation of niosomes by the scRPE method. These required conditions for the present PG ester-type nonionic surfactants are almost same as those for ethylene oxide-type nonionic surfactants reported in our previous study[19].

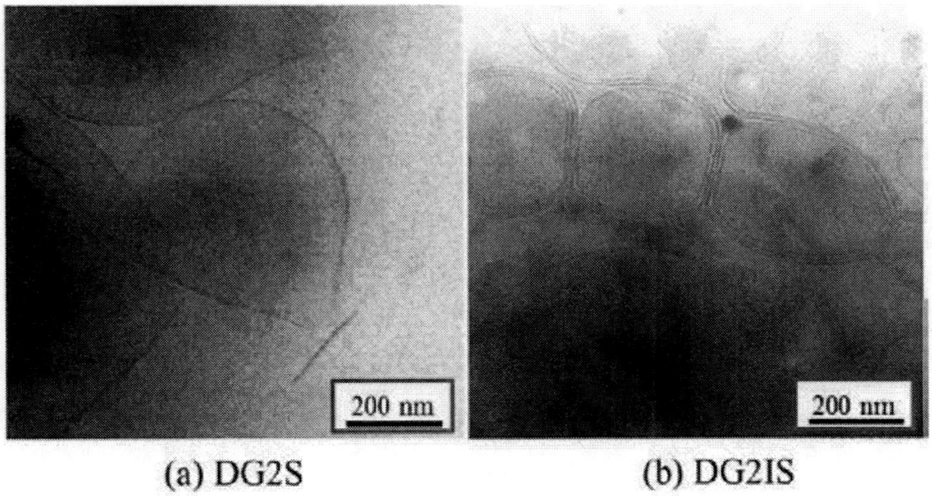

Figure 3. Cryo-TEM micrographs for (a) DG2S and (b) DG2IS niosome suspensions (10mM)

3.3. Comparison between DG2S and DG2IS niosomes

Fig. 4 shows average particle sizes of DG2S and DG2IS niosomes as a function of time after the sample preparation. The particle size of DG2IS niosomes kept an almost constant value at least for 1 week, while that of DG2S niosomes grew gradually from 350 to 1500 nm. This result suggested that DG2IS niosomes had higher dispersion stability than DG2S niosomes. Fig. 5 shows DSC curves for DG2S and DG2IS niosome suspensions. An endothermic peak was observed at 42°C for the DG2S niosome suspension, which indicated a phase transition from gel state (L_β) to liquid crystalline state (L_α). On the other hands, there was no endothermic peak for the DG2IS niosome suspension in the temperature range from 25°C to 70°C. Fig. 6 shows fluorescent anisotropy of DPH incorporated in DG2S and DG2IS niosomes as a function of temperature. The fluorescent anisotropy in DG2S niosomes was gradually decreased from 0.14 to 0.12 up to 40°C, and then rapidly decreased and reached to 0.08 above 45°C. In the case of the DG2IS niosome suspension, the fluorescent anisotropy showed an almost constant value of 0.07 in the temperature range from 5°C to 50°C. These results, as well as the DSC results, showed that the membrane fluidity of DG2IS niosomes was higher than that of DG2S niosomes at 25°C. The cryo-TEM image for DG2S niosomes (Fig.3a) was non-spherical and angular shape having some corners, which also suggested that these niosomes consisted of less flexible bilayers. While niosomes formed by DG2IS with higher membrane fluidity were almost spherical (cryo-TEM image in Fig. 3b). This higher membrane fluidity gives the DG2IS niosomes with higher dispersion stability and higher trapping efficiency.

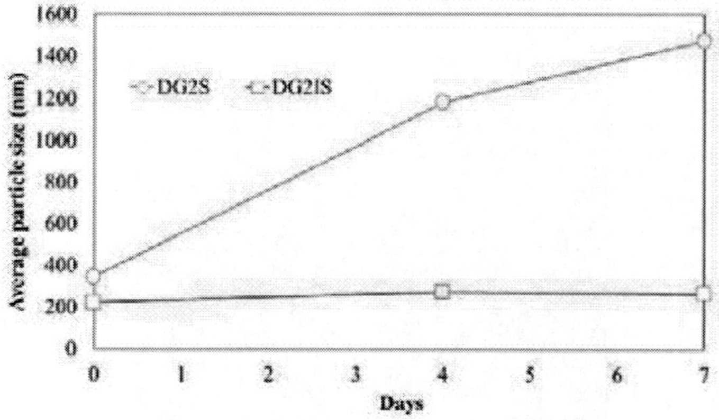

Figure 4. Average particle sizes obtained by DLS for DG2S and DG2IS niosome suspensions as a function of time after the sample preparation

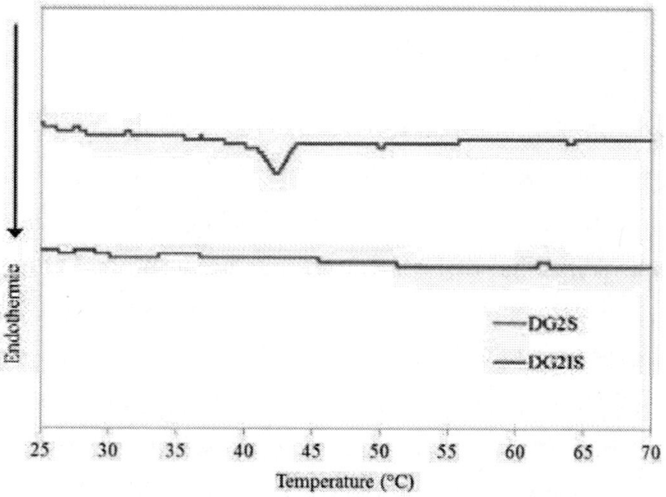

Figure 5. DSC curves for DG2S and DG2IS niosome suspensions.

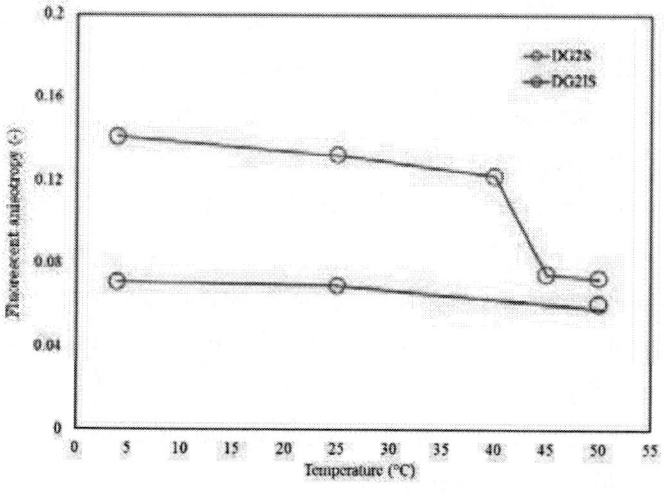

Figure 6. Fluorescent anisotropy for DG2S and DG2IS niosome suspensions as a function of temperature.

4. Conclusions

In this study, we prepared niosomes using PG ester-type nonionic surfactants by the scRPE method. The addition of ethanol as a co-solvent was needed for complete dissolution of the PG ester-type surfactants in $scCO_2$. For DG2S and DG2IS, we succeeded in the formation of niosomes by the scRPE method in the presence of 7.5 % ethanol. These nonionic surfactants have proper HLB values to form lamellar structure in aqueous solutions, which is one of the important factors to obtain niosomes in the scRPE method. Niosomes prepared with DG2IS have higher trapping efficiency and higher dispersion stability than DG2S niosomes because of higher membrane fluidity of DG2IS arisen from branched hydrophobic moiety in the molecule. These findings obtained in this study give us useful information to prepare niosomes using nonionic surfactants derived from naturally sources by the present environment-friendly method.

References

1) Bangham, A. D.; Standish, M. M.; Watkins, J. C. Diffusion of univalent ions across the lamellae of swollen phospholipids, *J. Mol. Biol.*, **13**, 238-252 (1965).

2) Balon, K.; Riebesehl, B. U.; Mueller, B. W. Determination of liposome partitioning of ionizable drugs by titration, *J. Pharm. Sci.*, **88(8)**, 802-806 (1999)

3) Colic, M.; Morse, D. The elusive mechanism of the magnetic `memory' of water, *Colloids Surf., A,* **154 (1-2)**, 167-174 (1999)

4) Montero,M.T.; Hernandez-Borrell, J.; Keough, K.M.W., Fluoroquinolone-biomembrane interactions: monolayer and caloriemtric studies, *Langmuir,* **14 (9)**, 2451-2454 (1998)

5) Komatsu, H.; Okada, S., Increased permeability of phase-separated liposomal membranes with mixtures of ethanol-induced interdigitated and non-interdigitated structures, *Biochim. Biophys. Acta,* **1237 (2)**, 165-175 (1995)

6) Imura T.; Yanagishita H.; Ohira J.; Sakai H.; Abe M.; Kitamoto D., Thermodynamically stable vesicle formation from glycolipid biosurfactant sponge phase, *Colloids and Surfaces B: Biointerfaces*, **43**, 115-121 (2005)

7) Yokouchi y.; Tsunoda Y.; Imura T.; Yamauchi H.; Yokoyama S.; Sakai H.; Abe M., Effect of adsorption of bovine serum albumin on liposomal membrane characteristics, *Colloids and Surfaces B: Biointerfaces*, **20**, 95-103 (2001)

8) Tianshun L.; Rodney J.Y.Ho, Trends and developments in liposome drug delivery systems, *J. Pharm. Sci.*, **90(6)**, 667-680 (2001)

9) Kaneda Y., Virosomes: evolution of the liposome as a targeted drug delivery system, *Adv. Drug Delivery Rev.*, **43 (2-3)**, 197-205 (2000)

10) Marjan, J.; Xie, Z.; Devine, D. V., Liposome-induced activation of the classical complement pathway does not require immunoglobulin, *Biochim. Biophys. Acta,* **1192 (1)**, 35-44 (1994)

11) Manosroi A.; Wongtrakul P.; Manosroi J.; Sakai H.; Sugawara F.; Yuasa M.; Abe M., Characterization of vesicles prepared with various non-ionic surfactants mixed with cholesterol, *Colloids Surf. B: Biointerfaces*, **30**, 129-138 (2003)

12) Nasseri B., Effect of cholesterol and temperature on the elastic properties of niosomal membranes, *Inter. J. Pharm.*, **300**, 95-101 (2005)

13) Uchegbu I.F.; Vyas S.P., Non-ionic surfactant based vesicles (niosomes) in drug delivery, *Inter. J. Pharm.*, **172**, 33-70 (1998)

14) Batzru S.; Korn E.D., Single bilayer liposomes prepared without sonication, *Biochim. Biophys. Acta*, **298**, 1015-1019 (1973)

15) Szoka F. Jr.; Papahadjopoulos D., Procedure for preparation of liposomes with large internal aqueous space and high capture by reverse-phase evaporation, *Proc. Natl. Acad. Sci. U.S.A.*, **75**, 4194-4189 (1978)

16) Otake K.; Imura T.; Sakai H.; Abe M. Development of a New Preparation Method of Liposomes Using Supercritical Carbon Dioxide, *Langmuir*, **17**, 3898-3901 (2001).

17) Otake K.; Shimomura T.; Goto T.; Imura T.; Furuya T.; Yoda S.; Takebayashi Y.; Sakai H.; Abe M., Preparation of Liposomes Using an Improved Supercritical Reverse Phase Evaporation Method, *Langmuir*, **22(6)**, 2543-2550 (2006)

18) Ri K.; Yamaguchi S.; Wongtrakul P.; Hashimoto S.; Otake K.; Ohkubo T.; Sakai H.; Abe M., Preparation and Characterization of Nonionic Surfactant Vesicles Using Supercritical Carbon Dioxide, *Mater. Technol.*, **23(5)**, 340-347 (2005)

19) Yamaguchi S.; Tsuchiya K.; Abe M.; Sakai H., Preparation and Solution Properties of Nonionic Vesicles Using Super Critical Carbon Dioxide Reverse Phase Evaporation Method, *J. Oleo Sci.*, submitted (2015)

20) Miwa, I.; Okuda, J.; Meada, K.; Okuda, G., Mutarotase effect on colorimetric determination of blood glucose with -D-glucose oxidase, *Clin. Chim. Acta,* **37**, 538-540 (1972)

21) Okuda, J.; Miwa, I., Enzymatic micro-determination of D-glucose and its anomer, *Protein, Nucleic Acid, Enzyme,* **17**, 216-224 (1972)

474177 Poloxamer-Based Formulations: Function through Molecular Self-Assembly and Directed Assembly

Abstract:

Amphiphilic polymers of the poly(ethylene oxide)-block-poly(propylene oxide) (PEO-PPO) family, commercially available as Pluronics or Poloxamers and approved for pharmaceutical use [Gu & Alexandridis, J. Pharm. Sci. 2004, 93 (6), 1454-1470; DOI: 10.1002/jps.20021], offer prime examples of self-assembling systems that find diverse applications. The presentation will utilize research findings from our group and others to highlight how the fundamental aspects of Poloxamer block copolymer micellization [Kaizu & Alexandridis, Colloids Surfaces A 2015, 480, 203-213; DOI: 10.1016/j.colsurfa.2014.10.061] and lyotropic liquid crystal formation and nanostructure [Sarkar et al., Macromol. Chem. Phys. 2012, 213 (23), 2514-2528; DOI: 10.1002/macp.201200438] in aqueous media, inform the formulation of Poloxamers applied to emulsions [Kaizu & Alexandridis, J. Colloid Interface Sci. 2016, 466, 138-149; DOI: 10.1016/j.jcis.2015.10.016], colloidal dispersions [Bodratti et al., J. Dispersion Sci. Tech. 2015, 36 (12), 1806-1815; DOI: 10.1080/01932691.2015.1011273], nanoparticle synthesis [Sakai et al., Colloids Surfaces A 2015, 487, 84-91; DOI: 10.1016/j.colsurfa.2015.09.058], and drug delivery [Zhang et al., Nature Commun. 2016, 7, DOI: 10.1038/ncomms11649].

463024 Chemical Engineering Body of Knowledge: An Academic Perspective

Abstract:

Chemical Engineering is increasingly multidisciplinary. Chemical Engineers work in fields ranging from healthcare, pharmaceuticals, oil and gas, chemicals and plastics to electronics and advanced materials. Driven by current trends in research funding, chemical engineering education and research has tended to emphasize molecular level processes. However recent studies show that employers of chemical engineering undergraduates want students trained in traditional areas with a strong understanding of macroscopic processes. Thus a gap is developing between the training received by undergraduates and the expectation of industry.

The Chemical Engineering Body of Knowledge provides a framework that highlights the areas of competence expected by employers of chemical engineering graduates. The Body of Knowledge could be a very useful tool when designing undergraduate courses. In this presentation some of the discrepancies between the education and training received by undergraduates versus the expectations of employers will be discussed. The potential role of the Chemical Engineering Body of Knowledge in bridging this gap will be highlighted.

473571 *Use of Bok By an AIChE Working Group in the Development of a New Process Safety Management (PSM) Curriculum for Aiche Academy*

Abstract:

Learn how the "Body of Knowledge for Chemical Engineers (BOK)" was used by an AIChE working group in the development of a new Process Safety Management (PSM) curriculum for AIChE Academy. This practical example shows how BOK engineering roles and Knowledge, Skills and Ability (KSA) matrices were used to create a highly visual, self-directed learning guide, which is being incorporated into the AIChE Academy's new PSM curriculum.

453445 Future Direction of BOK Development

Abstract:

There is no doubt that Body of Knowledge (BOK) published by AIChE will become the core of all the future academic and industrial activities in the field of chemical engineering. However, the establishment of BOK is just a starting point: BOK is not a BOOK; instead, it should be a system with a close loop of PDCA. The dedicated team should be set up to complete these regular and necessary tasks so that a dynamic BOK can benefit both the stakeholder and the audience. The presentation aims to provide a proposal of such kind of systematic approach and it is hoped to help the development of BOK in the future.

476740 Biotechnology in the Era of the Fourth Industrial Revolution

Abstract:

We are observing a big paradigm shift in how we live, work and even play based on such unprecedentedly fast development of various technologies. This tide of change is so big, and thus is called â€œthe Fourth Industrial Revolutionâ€. It is difficult to predict how this revolution will affect our life, but it is clear that it is already here with us. Integration of digital-physical-bio technologies is driving the fourth industrial revolution. It is best to understand the impact of the fourth industrial revolution through linking core developments with the worldâ€™s most important pending problems. Biotechnology will play increasingly important roles in producing what we eat and what we use in sustainable manner. Also it will become more and more important in improving our health and healthcare system. In this lecture, I will describe briefly how the fourth industrial revolution is being unfold and what biotechnology can deliver in the era of the fourth industrial revolution. Examples will be given on sustainable production of chemicals, fuels and materials through biotechnology and on how to improve our healthcare system in the aging society with increasing population.

474526 Invited Presentation By 2016 Management Award Recipient--Enriching the Future of East Tennessee through Environmental Clean up and Stewardship

Abstract:

As the recipient of the 2016 AIChE Management Division Award, Ken Rueter is living testimony to the fact that chemical engineering can be a rewarding career that offers cutting edge opportunities in innovation and problem solving. His experience underscores the value of the chemical engineering profession in addressing unprecedented national technical challenges.

As President and Project Manager of URS|CH2m Oak Ridge LLC (UCOR), Rueter has led his company's success in providing positive solutions for large-scale nuclear environmental cleanup, restoring once contaminated land for future economic development purposes.

While most engineers win acclaim for the enduring projects they design and build, Ken Rueter's most recent noteworthy achievements stem from what he and his company are demolishing and cleaning up. Today, across the U.S. Department of Energy (DOE) complex, millions of gallons of liquid radioactive waste are being processed. Thousands of tons of spent nuclear fuel and special nuclear material are being disposed. Huge quantities of contaminated soil and water are being remediated. And thousands of excess facilities are being deactivated and demolished.

This massive environmental cleanup – involving some of the most dangerous materials known to humankind – results from over five decades of nuclear weapons development and production and Government-sponsored nuclear energy research. In Oak Ridge, this work addresses the legacy of the super-secret World War II Manhattan Project where uranium was enriched for use in the atomic weapons that ended the war and, later, for fuel in commercial nuclear power plants.

In 2016, UCOR completed demolition and cleanup of the gaseous diffusion facilities at the East Tennessee Technology Park in Oak Ridge, Tennessee. This was a historic moment for a community that carved its place in history during World War II and the Cold War. It marked the first time in the world that all of a site's uranium enrichment process buildings were removed anywhere, eliminating environmental hazards and paving the way for future economic development. In Oak Ridge, that task included deactivating and dismantling 4.5 million square feet of radiological and chemically contaminated facilities and shipping more than 46,000 truckloads of debris.

These facilities include the massive and historic government uranium enrichment facilities from the Manhattan Project and the Cold War eras which played a key role in ending World War II and keeping the nation safe and energy-sufficient for decades afterwards. One of those facilities – the once super-secret K-25 gaseous diffusion plant – consisted of over two million square feet and in its time was the largest building under one roof in the world.

UCOR is working closely with its multiple partners in government, labor and the private sector to accomplish this massive environmental clean up and remediation project. UCOR's partnership with the U.S. Department of Energy and its labor force, including union crafts, is unprecedented in the nuclear industry. With its collective technical competency, UCOR has accomplished a cleanup task that others will never have the opportunity to address.

This achievement brings the Department of Energy closer to its ultimate vision for the site, now known as the East Tennessee Technology Park, as a privately owned and operated industrial park and home to new job-producing industries.

UCOR has completed these first-of-a-kind demolition tasks involving some of the most hazardous materials known to man with an unprecedented safety record – amassing more than 7 ½ million hours with a Lost Day Away accident and being recognized by DOE with Voluntary Protection Program Star status, recognizing the safest worksites in the nation.

UCOR, like one of its parent companies AECOM, is built to deliver a better world. UCOR takes its inspiration from AECOM's transformative work in designing, building, financing and operating infrastructure assets for governments, businesses and organizations in more than 150 countries. As a fully integrated firm, AECOM connects knowledge and experience across its global network of experts to help clients solve their most complex challenges – from building high-performance buildings and infrastructure to creating resilient communities and environments, like the revitalized East Tennessee Technology Park in Oak Ridge.

458045 Your Career Is a Startup

Abstract:

As chemical engineers, we are equipped with highly sought after skills to tackle technological and strategic problems. But how do we apply those same skills towards the development of our own careers? What if your career was run like a startup? What would be your assets? What customers would you serve? What would make you the go-to person when certain challenges arise? How would you market yourself to your company and to your profession?

Participants will learn:

1. The difference between personal brand identity vs. brand equity
2. How to leverage those technical skills youâ€™re already great at to build your brand
3. Ways to practice executive presence, from how you hold face-to-face meetings with your manager, to presenting â€œYouâ€ in your best light to the executive levels of your company
4. How to use your skills as an engineer to analyze your current career path

475779 Chemical Engineering Careers: Community Development Gaps

Abstract:

This talk will cover three topics that I view as gaps along Chemical Engineering career paths that need volunteer support, informational resources, and collaborative address. **First:** Focus on Chemistry and Chemical Engineering. The two professions are classically separate, but in the future have the potential to innovate with far and deep reaching impact. A few examples of how practitioners of holistic design that have an impact on our world will be shared. **Second:** I will focus on being more than a chemical engineer, a chemical engineer plus some if you will (*i.e.,* ChE + S). In the past I have spoken to soft skill development but in this talk will illustrate how multidisciplinary has become a new norm with examples from my own experiences. One example for instance, is the institute has suggested that credentialing is important. While there is a push for Professional Engineering licensure, there are a myriad of accreditation processes that can aid ChEs on their career paths. **Third**: expectations for leadership and implementing change are different than they used to be. Instituting change requires standing against odds and making a difference on many scales and under adverse circumstances. Focus on this third topic will be given to the development of resources in education settings for chemical engineers.

This talk will demonstrate diversity among career paths that still require ChE community support. This talk is geared to promote collaborative discussion and action in progressive career development, please attend if you would like to support developing these topics within and for the Chemical Engineering community.

453874 Molecular Modeling of Materials: A Chemical Engineerâ€™s Perspective

Abstract:

Since the 1950â€™s John Prausnitz has brought into Chemical Engineering key concepts and methods of physical chemistry, enabling the prediction of thermodynamic properties and phase equilibria in a wide variety of multicomponent mixtures. His work has had tremendous impact on industrial process and product design. Inspired by his contributions, molecular-based approaches for understanding and tailoring structure-property-processing relations in materials, based on the fundamental principles of quantum and statistical mechanics, have gained ground in chemical engineering research and practice. They have been greatly aided by an unprecedented growth in computer power, but also by new, efficient methods and algorithms, in whose development chemical engineers have played a pivotal role. The broad length and time scales governing structure and dynamics in real-life materials have demanded the advancement of multiscale modeling strategies, involving more than one levels of representation, to bridge atomistic constitution and interactions with macroscopic properties. This is very much in line with the synthetic, problem-oriented thinking of chemical engineers.

In this talk we will discuss examples of molecular modeling of polymeric and nanostructured materials that are relevant to the automotive and microelectronics industries: how can we push the frontiers of predictability by appropriate design of multiscale theoretical and simulation approaches and what can we learn about processing and end-use properties?

CONTRA COSTA
HEALTH SERVICES

Process Safety Management Now and in the Future

Celebrating NorCal's 70th Anniversary

Randy Sawyer

What is Process Safety

- Process Safety is defined as "a discipline that focuses on the prevention of fires, explosions, and accidental chemical releases at chemical process facilities"
- Is the prevention of, preparedness for, mitigation of, response to, or restoration of catastrophic releases of chemicals or energy from a process associated with a facility.

CONTRA COSTA
HEALTH SERVICES

Process Safety Management History

- Nitroglycerine
- Qualitative Analyses
  - Hazard and Operability Studies (HazOp)
  - Failure and Effect Mode Analysis (FMEA)
  - Checklists/What if
- Quantitative Analyses
  - Fault Tree Analysis (FTA)
  - Quantitative Risk Assessments (QRA)
- Layer of Protection Analysis (LOPA)

CONTRA COSTA
HEALTH SERVICES

Process Safety Accidents that Defined Process Safety

- Flixborough – Cyclohexane released from a temporary bypass – 28 workers died, 26 injured and 53 offsite injuries
- Seveso – Runaway reaction that released dioxin into the residential area – many people developed Chloracne and 6.6 square miles was made inhabitable

THE GAS CLOUD

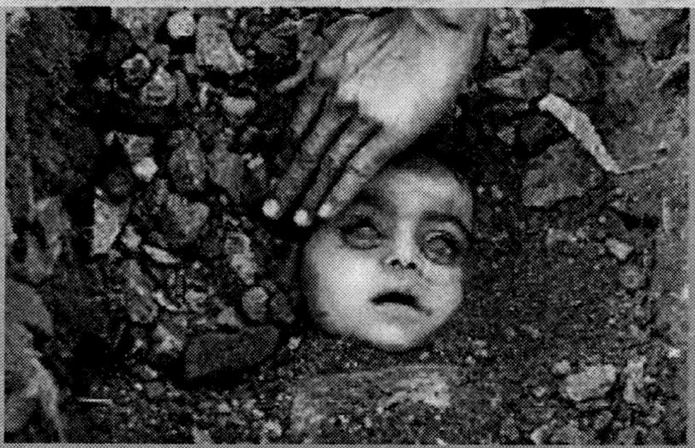

Process Safety Accidents that Defined Process Safety

- Bhopal – Release of methyl isocyanate where over 3,000 people died and many more were injured

Process Safety Management Regulations

- Process Safety Management is a management system that includes twelve elements:
  - PSI
  - PHA
  - Operating Procedures
  - Training
  - Mechanical Integrity
  - Management of Change
  - Pre Startup Safety Reviews
  - Contractor Safety
  - Incident Investigation
  - Compliance Audits
  - Hot Work
  - Employee Participation

CONTRA COSTA
HEALTH SERVICES

Contra Costa County Industrial Safety Ordinance

- 1990's Major Incidents
- County's Board of Supervisors and Public Concerns led to the Industrial Safety Ordinance
- Expands on the California Accidental Release Prevention Program
  - Inherent Safety
  - Human Factors
  - Root Cause Analysis
  - Management of Organizational Change
  - Safety Culture Assessments
  - Safeguard Protection Analysis
  - Process Safety Indicators

CONTRA COSTA
HEALTH SERVICES

Accidents are Still Occurring

- BP Texas City Refinery
- Deep Water Horizon
- Tesoro Anacortes Refinery
- Chevron Richmond Refinery
- West Texas

CONTRA COSTA
HEALTH SERVICES

What is Happening Now

- California Accidental Release Prevention Program 4 and PSM Regulations
- US EPA proposed changes to the Risk Management Program
- OSHA will be proposing changes to the Process Safety Management Regulations
- The State of Washington considering expanding regulations for petroleum refineries

CONTRA COSTA HEALTH SERVICES

Where Do We Go Now?

- CCPS Risk Based Process Safety
- Vision 20/20
- CCPS Certified Process Safety Professional
- Process Safety Culture

CONTRA COSTA HEALTH SERVICES

Risk Based Process Safety

- Commit to Process Safety
  - Process Safety Culture
  - Compliance with Standards
  - Process Safety Competency
  - Workforce Involvement
  - Stakeholder Outreach
- Understand Hazards and Risk
  - Process Safety Knowledge Management
  - Hazard Identification and Risk Analysis

CONTRA COSTA
HEALTH SERVICES

Risk Based Process Safety

- Manage Risk
  - Operating Procedures
  - Safe Work Practices
  - Asset Integrity and Reliability
  - Contractor Management
  - Training and Performance Assurance
  - Management of Change
  - Operational Readiness
  - Conduct of Operations
  - Emergency Management
- Learn from Experience
  - Incident Investigation
  - Measure and Metrics
  - Auditing and Management Review
  - Continuous Improvement

CONTRA COSTA
HEALTH SERVICES

VISION 20/20

Five Company Tenets and Four societal Themes

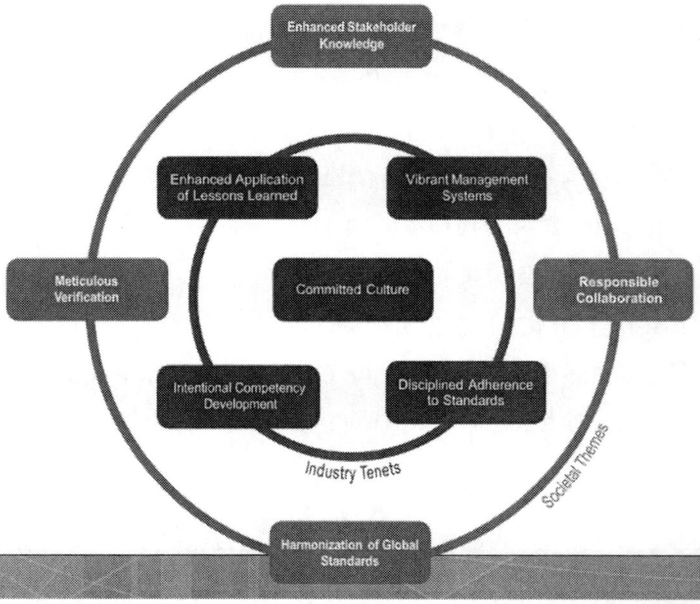

CCPS Certified Process Safety Professional

- CCPS Certifying Process Safety Professional go to: http://www.aiche.org/ccps/resources/certified-process-safety-professional

CONTRA COSTA HEALTH SERVICES

Process Safety Culture

- Management Commitment
- Employee Involvement
- Assessments
- Address Areas for Improvement
- Reassess to determine if improvement plan is making a difference
- New Assessment address new areas of concern
- Continue

CONTRA COSTA
HEALTH SERVICES

- Randy Sawyer
- Chief Environmental Health and Hazardous Materials Officer
- Randy.sawyer@hsd.cccounty.us
- 925-335-3210

CONTRA COSTA
HEALTH SERVICES

475521 Microreaction technology and applications: 3-printed microreactors

Abstract:

Continuous-flow microreaction technology has exhibited intrinsic advantages of reduced chemical consumption, design for energy efficiency, and inherently safe chemistry. In here, I present various microreactors and their applications that have been achieved at our laboratory, as long as time is allowed

1. Automated Chemical Process

- A total process concept was applied to carcinogenic, explosive, toxic, or noxious regents (diazomethane, isocyanides, diazoacetate, chloromethyl methyl ether) by serially conducting the generation, its purification, its utilization for a desired synthesis.
- Various microseparations on the basis of different principles such as liquid-liquid extraction, liquid-gas distillation, membrane separation in purifying the generated reagents were demonstrated.
- Unstable intermediate with short-life time was generated and reacted with a infused reagent, which performed within 330 Î¼sec for outpacing Fries rearrangement through microfluidic rapid mixing

2. 3D Printed Ceramic Microreactor

The versatile structures of 3D-printed preceramic monoliths with 1~11 cm scale were fabricated from a modified polyvinylsilazane resin as a novel single precursor, then transformed to silicon carbonitride (SiCN) ceramic bodies upon pyrolytic conversion to 1000 °C. The fully dense SiC-based ceramic microreactor with excellence in thermal and chemical stability was extended to devise a hydrogen production system by decomposition of corrosive ammonia gas at high temperatures upto 1000 °C

475522 Nanomanufacturing of Functionalized Carbon Electrodes for High-performance Energy Storage Devices

Abstract:

High-performance electrochemical energy storage (EES) systems can increase the efficiency of new transportation and renewable energy storage technologies. Among EES systems, rechargeable Li-ion battery has become the most popular power source for consumer electronics and electric propulsion systems due to its high energy density and efficiency. However, the high cost of its cathode materials, such as inorganic transition metal oxides, limits installations of batteries for large-scale energy storage applications, including renewable energy storage and electric grid applications.

Therefore, the development of more sustainable electrode materials is a critical challenge to support the fast expansion of batteries for large-scale energy storage applications. Recently, sustainable organic electrode materials composed of earth abundant elements have been intensively investigated to replace inorganic cathode materials in Li-ion batteries. We also investigate charge storage characteristics of functionalized carbon materials, including oxidized carbon nanotubes, reduced graphene oxides, biomass derived carbons, and small organic molecules. We demonstrate 3D nanostructured electrodes using these functionalized carbon materials via various nanomanufacturing processes. The assembled functionalized carbon electrodes are employed as cathodes in Li-cells, delivering high capacity with superior power capability and cycling stability. Thus, these high-performance functionalized carbon electrodes can be promising cathodes for large-scale rechargeable batteries or hybrid capacitors.

475523 Diffusion-mediated Photolithography for Designing 3D Microstructures

Abstract:

Photolithography has long been used for the preparation of polymeric micro- and nano-patterns in various fields of industry, including semiconductor processing. In general, the directionality of light propagation yields only two-dimensional (2D) patterns through conventional photolithography. We have developed a novel photolithographic approach based on reaction-diffusion process to create 3D patterns without the use of complex optical setup or special materials. During free-radical polymerization, oxygen rapidly consumes radical to form peroxide, thereby inhibiting polymerization. Therefore, the spatial distribution of oxygen concentration can significantly influence the growth of polymeric structure during ultraviolet (UV) exposure. When the rate of oxygen-inhibition reaction is comparable with that of oxygen diffusion, the oxygen gradient is spontaneously formed in the UV-exposed region due to the influx of oxygen from the unexposed region. Moreover, the gradient can be further engineered by employing different sets of boundary conditions for oxygen diffusion. Therefore, growth pathways of polymeric structures can be selected by controlling the boundary conditions, one of which enables the creation of microparticles and patterns with unconventional 3D shapes. Moreover, stepwise selection and application of distinct pathways further develop the microstructures to have complex shapes. The ease of use and high controllability of this technology will provide new opportunities for various applications of polymeric structures, including wettability-controlled surfaces and smart bioactive carriers.

475524 Integrated Fabrication-Conjugation Approaches for Tunable Manufacturing of Biofunctional Hydrogel Microparticles

Abstract:

Hydrogel microparticles with controlled structures have gained increasing attention in various applications as biosensing platforms for medical diagnostics, biological threat detection and bioprocess monitoring. However, it still remains challenging to manufacture such microparticles with tunable structures and multifunctionality and in a simple, cost-efficient and reliable manner. We present facile manufacturing methods for chemically functional and monodisperse biopolymeric–synthetic hybrid hydrogel microspheres with controlled macroporous structures that can be utilized as conjugation platforms for large biomolecules. Specifically, we exploited a simple micromolding-based approach in order to fabricate highly uniform poly(ethylene glycol) (PEG)- and poly(acrylamide) (PAAm)-based hydrogel microspheres and shape-encoded microparticles, in which a potent aminopolysaccharide chitosan providing chemically reactive and abundant amine groups is incorporated. The as-prepared microparticles showed macroporous and/or intriguing core-shell structures stemming from either spontaneous polymerization-induced phase separation or via easy porosity tuning with inert PEG porogens. Fluorescent labeling studies utilizing an amidation reaction on the chitosan's primary amines confirmed stable incorporation of the chitosan moieties with retained chemical reactivity, showing potential of such amines as efficient conjugation handles. The utility of our hydrogel microparticles as biomolecular conjugation platforms was thoroughly examined in terms of conjugation capacity and kinetics by conjugating model biomolecules (fluorescein-labeled single-stranded DNA and red fluorescent protein R-phycoerythrin) via two high-yield bioorthogonal reactions (strain-promoted alkyne–azide cycloaddition (SPAAC) and tetrazine–***trans***-cyclooctene (Tz–TCO) cycloaddition). The conjugation results showed well-defined and macroporous network structures, leading to enhanced protein conjugation capacity and kinetics and allowing programmable protein conjugation. In this presentation, our recent progress on much improved macropore formation and conjugation for the PAAm-based hydrogel microparticle with acrylate functionality via EDC/NHS chemistry will also be highlighted.

475525 A New Approach for Ultra-thin Tunable Molecular Sieve Membranes for High-resolution Custom Gas Separations

Abstract:

Molecular-sieving zeolitic imidazolate frameworks (ZIFs), a sub-class of metal-organic frameworks (MOFs), offer unique opportunities in gas separations primarily due to their ultra-micropores (pores smaller than 5 Å) and their unusual thermal/chemical stabilities. However, there exist both fundamental materials limitations and processing challenges: one major fundamental materials challenge is that only very few ZIF/MOF materials show promising molecular sieving properties for industrially important gas separations. The reason is that an exact one-to-one size match between the aperture size and the molecules of interest is required. However, as with any other crystalline porous materials such as zeolites, when the size disparity of the gas molecules of interest is too small as is the case for many challenging separations, ZIFs (MOFs in general) with suitable aperture size are not always available. Processing challenges include slow batch crystallization, grain boundary defects, expensive porous supports, lack of "generalâ€ method, and reproducibility issues. Though some of these processing challenges have been recently addressed, there still remains an ultimate challenge for the practical applications of ZIF membranes, which is the high cost of membranes and membrane processes as compared to polymeric membranes. One can bring down the cost of membranes and membrane processes by significantly improving the productivity of membranes by 1) substantially reducing membrane thickness and 2) increasing membrane area (note that the membrane productivity is given as the following equation: , where P_A, l, and A are the permeability of gas A, membrane thickness, and membrane area, respectively).

In this talk, I would like to discuss a completely new approach to address both materials and processing challenges mentioned above. The approach is based on the labile coordination chemistry of metal-organic framework materials, enabling to post-synthetically design, engineer, and modify the fine structures of ZIFs such that the aperture sizes of ZIFs can be finely tuned for custom gas separations as well as the thicknesses of their membranes can be in sub 10 nanometer range.

475663 Application of Field-Flow Fractionation to the Shape-based Separation of Rod-like Particles: Approach by Theoretical/Numerical Models

Abstract:

Field flow fractionation (FFF) has been applied to separate different size and shape of nanoparticles. Most of recent studies are focused on FFF of sphere particles while many applications are also based on separation of other particle shapes such as rod like nanoparticles. In this study various mathematical models have been derived and studied to better explain rod-like nanoparticle behavior in a field flow, also, it was assumed that rotational diffusivity is much higher than the transverse diffusion of the rod, therefore throughout this work, average local diffusivity of the particle was calculated based on a position dependent orientation average. A model was proposed to evaluate the concentration of particles along the cross-section which could be translated into a retention ratio of the nanoparticle. It was shown that the hydrodynamic interaction with wall known as "lift-hyperlayer" mode has an influence on the migration rate of the rods. Also it was demonstrated that despite previous studies, rod transvers diffusivity differs from its perpendicular diffusivity when it is not assumed to be always aligned in the axial direction. To better understanding the results, retention rate of the rod like particles was used to compare with sphere particles in several operational conditions. In another approach, the "pole-vault" behavior of rods near the wall was to taken into consideration to derive an improved prediction of the steric-entropic effect. This effect basically is a result of the inability of rod to have its own full rotation due to interactions with wall. As demonstrated in this paper, the aspect ratio of rod affects particle distribution along the cross-section, subsequently the retention ratio of the nan rods. A comparison between the new model and the original model has been done in this paper. This model is being improved by experimental study in order to come into better agreement with experimental results.

467843 What Is the International Impact of Europe's Bologna Process?

Abstract:

In 1999, Education Ministers from 29 European countries signed the Bologna declaration, creating the European Higher Education Area. What is the impact on Chemical Engineering education? What does it mean to internationalization in our field? Official statements, independent assesments, student's views will be compiled. The paper can be discussed with the audience.

460817 Student-Created Assessment: Midterm Test of a Graduate Course, Mathematical Methods in Chemical Engineering

Abstract:

Students were divided into groups, each of which was asked to provide a question paper and suggested solutions like a course instructor, as their midterm assessment. The course, Mathematical Methods in Chemical Engineering, is a graduate core course, mainly covering analytical and numerical techniques for solving differential equations generated by modeling of chemical processes. The class size was normally between 30 and 50. Detailed guidelines and sample product created by the instructor or previous students were provided at the beginning of the semester. The grading was based on: whether the questions in the product can assess student achievements of module learning objectives; whether a "normal" student can complete the paper reasonably well within 4 hours; the clarity of the problem statements; the significance of the physical problems; the reasonability of modelling and the correctness of using techniques to solve; the creativity or critical discussion on the problem, modelling, and/or the solving the model. An oral presentation, highlighting their ideas, was given by each group, for sharing and grading fairness. A survey was conducted at least 6 months after this assessment to get student opinions on this experiment. The product showed that most students achieved and knew how to assess the achievement of the module learning outcomes. Many creative questions and critical discussion on the questions came out, suggesting that students challenged their comfortable zone, and also learned through the assessment procedure. Students' responses to the survey questions showed their interests and acceptance of such type of assessment. They also expressed that they liked this learning experience, and it was fairer than the conventional exam, because students can show their mastery of knowledge in their own way, without being afraid of misunderstanding the questions.

459654 Feedback- and Engagement-Enhanced Education: A Newly-Designed Senior Course, Chemical Product Design

Abstract:

A new technical elective course, *Chemical Product Design*, has recently been offered by the Department of Chemical and Biomolecular Engineering at the National University of Singapore. This course is targeted to senior students (undergraduate Year 4 or masterâ€™s) to equipped them with skill of designing high-value-added specialty chemical products (e.g., devices, medicine, personal-care products), as an supplement of traditional process design skill (for commodity products). It covers the basic methodology with illustrative examples from many areas (based on the textbook: *Chemical Product Design*, by Cussler and Moggridge, 2nd edition, 2011). The course involves active-learning lectures and student teaching (which is recorded for self-reflection and feedback/evaluation from the lecturer, teaching assistant and peers) so that students gain competence of thinking divergently and critically, and confidently solving open-ended problems through group discussion. Besides the technical aspects, the module also emphasizes engineering professionalism (including ethics and sustainability) and communication skills (by supervised group discussion and student teaching with feedback loop). After completion of the course, students are able to apply the basic methodology to design a product in group, write a professional report, teach others based on the learning objectives, and evaluate otherâ€™s teaching.

466727 Student Activation and Interaction through Tutorials and Adequate Scheduling

Abstract:

Introduction

Activating students and interacting with them are hot topics in education. In this respect, blended learning, entailing the combination of a number of pedagogic approaches, irrespective of learning technology used, represents a valuable approach [1]. The idea of blended learning is founded upon the notion that every student has an individual way of processing new theories. By offering a broad range of educational tools, the chance for a student to successfully grasp a new concept drastically increases [2-5]. Currently, e-learning is being hyped as the ultimate form of blended learning. However, this form is not always easily accessible and similar outcomes might be obtained through another form of blended learning, e.g., tutorials. Moreover, a critical assessment and reorganization of the course schedule is expected to already significantly enhance the students' motivation.

Kinetic Modelling and Simulation: starting situation

At Ghent University, Belgium, Kinetic Modelling and Simulation (KMS) is one of the core courses in the chemical engineering program. During this course, 20 to 30 students are confronted with a multidisciplinary field comprising mathematics, statistics, chemistry, engineering and informatics aiming at simulating and, hence, elucidating chemical kinetics. The KMS course spans a single semester, i.e., 12 weeks, and was constructed upon 5 different pillars: theory, examples, exercises, projects and feedback, see Figure 1. The theory, example and exercise sessions were given during the contact hours. In total, 8 ex cathedra theory sessions of 3 clock hours were scheduled in these 12 weeks to cover 5 topics: statistics (A), linear regression (B), non-linear regression (C), experimental design and model discrimination (D) and multi response regression (E). Exercises were only given on the topic of non-linear regression and were situated relatively far in time from the corresponding theory sessions, c.q., the theory on non-linear regression was given in weeks 4 and 5 while the corresponding exercises were held only in weeks 6, 7 and 9. During the latter weeks, the theory corresponding to other concepts such as experimental design was taught which did not promote the students' learning process. Three projects were given on three different topics, i.e., linear regression, non-linear regression and experimental design aiming at model discrimination, after which students got overall feedback. At the end of the course, 4 to 5 PhD students presented their work as an example for different aspects in kinetic modelling.

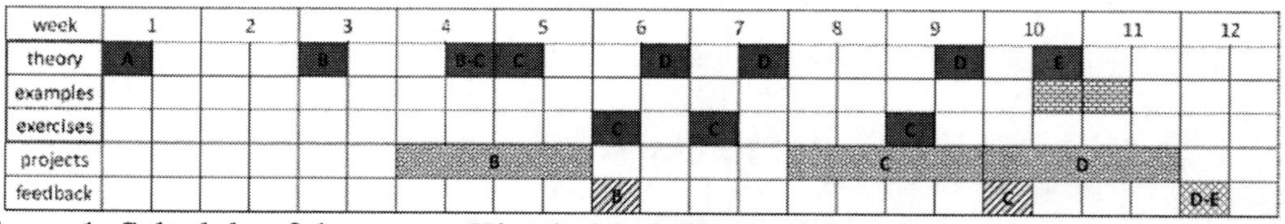

Figure 1: Schedule of the course Kinetic Modelling and Simulation before reorganization

Although the course has been evaluated rather positively over the years, some issues were identified: (1) during the theory sessions, **interaction was limited** due to the rather abstract content, (2) students were **ill-prepared** for the exercise sessions, (3) **not all concepts were put in practice** through exercises and/or projects and (4) throughout the semester, the **workload was concentrated at specific times**.

Reorganizing of the course according to blended learning principles

Aiming at enhancing the students' motivation and interaction with the teachers, a well-considered reorganization of the KMS course has been implemented, see Figure 2. Ex-cathedra theory blocks are, in this scheme, limited to 1.5 hours only and are followed by an equal amount of time for tutorials and hands-on trainings in which the students immediately process and apply the theory covered in the preceding time block. In addition, after a number of theory and tutorial sessions, a project assignment is given as a closure of that particular topic in which the student has to prove that he/she can apply the theory individually. After every project, which takes two weeks, students receive individual feedback. Compared to the previous schedule, the sequence of theory processing through ex-cathedra block, tutorials and projects is more continuous and, hence, contributes to a more blended learning environment. The work presented by the PhD students as kinetic modelling examples has been spread out more evenly at the end of the semester. Lastly, some additional topics have been introduced: contribution and reaction pathway analysis (F) and 'reading and analyzing a scientific publication' in regression analysis (G) which are well considered and relevant in the field of kinetic modelling.

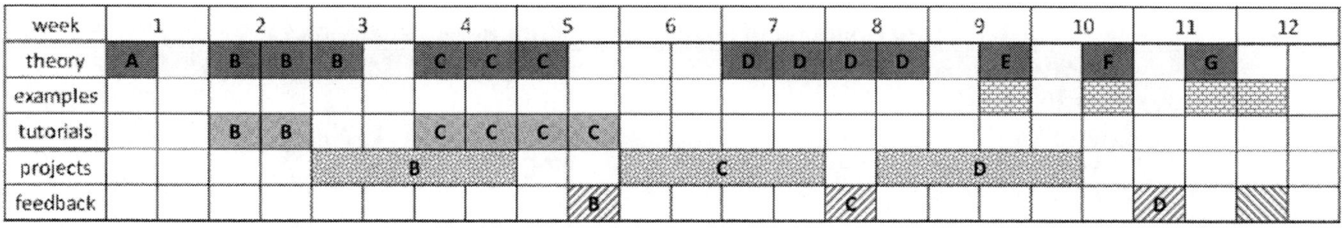

Figure 2: Schedule of the course Kinetic Modelling and Simulation after reorganization

Feedback and grading

Every student receives feedback in various ways throughout the semester. Firstly, all 6 tutorial sessions are led by a number of assistants, i.e., one assistant per 6 to 7 students. Hence, during these tutorials, every student has a close interaction with an experienced assistant. In addition, after every project, each student has personal feedback, not only on their hard skills, i.e., the project report content, but also on their soft skills, i.e., writing, visualizing and summarizing. At the end of the semester, there is a common feedback session in which the students give direct feedback to the teacher and assistants.

The final grade for the KMS course is composed of two parts. One third is earned via the three projects made during the semester. The remaining two thirds are obtained viva voce. Typically, this examination comprises a theoretical question and a case study or a scientific publication analysis. A student acquires the credits for the course when they obtain a weighted score exceeding 50% (10 out of 20) with minimum score of 40% on both components individually.

Benefits and challenges

An overview of the number of contact hours and sessions is given in Table 1. Concerning theory, the contact hours decreased from 24 to 22.5 hours while the number of contact sessions almost doubled. This yielded more time for the students in which they can process the concepts taught. Surprisingly, the time spent on exercises/tutorials stayed identical while the number of sessions increased with 100%. Since these sessions are planned directly after the corresponding theory session, the tutorials are tailored specifically to this theory. Finally, more time was allocated for the PhD students to explain their work within the framework of kinetic modelling to the students. Although it could already be implied from a comparison of Figures 1 and 2, this reorganization did not result in a significant increase in workload for the teacher and assistants. Moreover, the total number of contact hours remained equal while the number of contact sessions doubled. However, having a more strict schedule, requires the teacher and assistants to have a more strict schedule.

Table 1: Number of contact hours and sessions of the course Kinetic Modelling and Simulation before and after reorganization

	before reorganization		after reorganization	
	number of hours	number of sessions	number of hours	number of sessions
theory	24	8	22.5 (↘)	14 (↗)
exercises/tutorials	9	3	9 (=)	6 (↗)
examples	4.5	2	6 (↗)	4 (↗)
total	37.5	13	37.5 (=)	24 (↗)

Conclusions

By critically assessing and reorganizing the schedule for the course Kinetic Modelling and Simulation according to blended learning principles, the student/teacher/assistant interaction has significantly increased, as reflected by the increase in number of contact sessions. This resulted in a significantly augmented student motivation. Although the amount of time spent by the teacher and assistants during the course stays equal, it requires more flexibility. The immediate follow-up of theoretical by the corresponding tutorial sessions allows for a more direct learning process of the students, remediating their previous ill-preparedness for the exercises. In addition, the workload has been spread out more evenly over the semester.

References

[1] M. Oliver, K. Trigwell, Can 'Blended Learning' Be Redeemed, E-Learning, 2 (2005).

[2] N. Hoic-Bozic, V. Mornar, I. Boticki, A Blended Learning Approach to Course Design and Implementation, Ieee T Educ, 52 (2009) 19-30.

[3] M.V. Lopez-Perez, M.C. Perez-Lopez, L. Rodriguez-Ariza, Blended learning in higher education: Students' perceptions and their relation to outcomes, Comput Educ, 56 (2011) 818-826.

[4] V. Woltering, A. Herrler, K. Spitzer, C. Spreckelsen, Blended learning positively affects students' satisfaction and the role of the tutor in the problem-based learning process: results of a mixed-method evaluation, Adv Health Sci Educ, 14 (2009) 725-738.

[5] J.H. Wu, R.D. Tennyson, T.L. Hsia, A study of student satisfaction in a blended e-learning system environment, Comput Educ, 55 (2010) 155-164.

471386 An International Summer School for Undergraduate Chemical Engineers: The Imperial College Experience

Abstract:

The Imperial College Chemical Engineering Summer School programme is provided to international sophomore, junior and senior students studying chemical engineering and offers a schedule combining extensive laboratory experiments, a rig-building team exercise and a Pilot Plant project. The students are trained using a state-of-the-art carbon capture pilot plant which provides students with an experience that is as close to real industrial work as practical. As well as learning how to start up, stop and run efficiently pilot scale chemical engineering plant, the students become familiar with plant safety, shift handover protocols, permits to work, working is small shift teams, alarm management, PID diagrams and many other industrially relevant procedures. Summer school students will also take part in a bespoke range of short and long undergraduate laboratory experiments based on our current MEng. curriculum during their six week programme.

The paper will review the development of the summer school curriculum at Imperial College over the past 5 years, feedback from students and faculty from cohorts from 3 US colleges, and the future opportunities for summer schools as a way of enriching the studentâ€™s academic and cultural learning experiences. Comments on the use of UTA-undergraduate teaching assistants, group laboratory exercises and the overall importance of hands on teaching and learning will be discussed in the context of the current trend towards virtual and computer aided teaching. Â Â

459134 Global Engagement Strategies for Engineering Students at Western Michigan University

Abstract:
Global Engagement Strategies for Engineering Students at Western Michigan University

James Springstead, PhD, Assistant Professor

Said AbuBakr, PhD, Professor, AIChE Fellow

College of Engineering and Applied Sciences

Department of Chemical and Paper Engineering

Kalamazoo, MI

Abstract

Western Michigan University (WMU) has established global engagement as an important part of its vision and strategic objectives. To help implement these global strategic objectives, the faculty senate votes on a memorandum of action to encourage colleges to establish global learning outcomes. Instead of reinventing the wheel, the College of Engineering and Applied Sciences (CEAS) used already established ABET student outcomes that have global elements, with the WMU Department of Chemical and Paper Engineering serving as a leader in establishing clearly linked ABET performance indicators that are used to assess student outcomes.

Introduction

The visionÂ of Western Michigan University is:

Nationally and internationally recognized, the University aspires to distinguish itself as learner centered, discovery driven, and globally engaged.

Learner centered: Western Michigan University is a university where every member of our community is responsive to and responsible for the education of our students. We challenge and engage all members of our community with a university experience that creates skilled, lifeâ€long learners.

Â Discovery driven: Western Michigan University offers experiences that enable discovery, and promote creativity and research. We are committed to pursuing inquiry, disseminating knowledge, and fostering critical thinking that encourages lifeâ€long learning. Our scholarship creates new knowledge, forms a basis for innovative solutions, leads to economic development, and makes substantial contributions to society.

Globally engaged: Western Michigan University impacts the globe positively. We are a community of learners committed to human dignity, sustainability, social responsibility, and justice. Our campus embraces a diverse population of students, faculty and staff who develop learners and leaders who are locally oriented and globally competent, culturally aware and ready to contribute to world knowledge (1).

To implement this vision, WMU developed the following global objectives:

ObjectiveÂ 1.4:Â InternationalÂ learningÂ experiencesÂ prepareÂ allÂ learnersÂ forÂ theÂ globalizedÂ world.

Strategies

a. Increase feasibility of broadâ€based participation in study abroad programs and other globally focused learning opportunities.

b. Implement the facultyâ€adopted internationalized education initiative.

c. Provide enhanced livingâ€learning support for the growing number of international students at WMU.

d. Increase recruitment of University employees with international experience and expertise.

e. Facilitate intercultural understanding through greater interaction between domestic and international students.

ObjectiveÂ 2.1:Â WMUâ€™sÂ academicÂ programsÂ respondÂ toÂ studentÂ andÂ globalÂ needsÂ throughÂ innovative,Â highÂ qualityÂ curricula.

Strategies

a. Increase flexibility within the curriculum development process to better respond to a rapidly changing world.

b. Revise general education curricula to respond to the needs of the 21st century student.

Â c. Develop and utilize Universityâ€wide student learning outcomes for University planning and assessment.

d. Develop stronger connections across colleges and among external constituents to facilitate interdisciplinary opportunities.

e. Identify and support growth in the number of programs that achieve national or international distinction.

Objective 3.3: WMUâ€™s outreach is attentive to local and world needs.

Strategies

a. Develop and implement a renewed Campus Compact Action Plan, dedicating the University to work with other institutions to advance the public purposes of higher education.

b. Review and improve outreach practices and policies to achieve the Carnegie Community Engagement Reclassification by 2020.

c. Operationalize a unified vision to support the Universityâ€™s public engagement mission.

d. Develop rigorous, systematic evaluations of all outreach and community engagement programs to increase impact in conjunction with community agenciesâ€™ planning initiatives.

Â e. Know, measure, tell, and increase WMUâ€™s economic impact and social value.

f. Assist all students with developing meaningful connections in their local and global communities.

The international Education Council of Western Michigan University passed a resolution that was approved by the Faculty Senate asking colleges to create global Learning outcomes (GLOs)

The response of our college was that all our engineering, engineering technology, and computer science programs are ABET accredited. Global learning outcomes are already included in ABET student learning outcomes (3c), (3h), and (3j) and are periodically measured, assessed, and evaluated as required to maintain ABET accreditation (2).

Here are the ABET student outcomes that emphasize global engagements:

(c) an ability to design a system, component, or process to meet desired needs within realistic constraints such as economic, environmental, social, political, ethical, health and safety, manufacturability and sustainability.

The Department of Chemical and Paper Engineering measured metrics (performance indicators):

C3 Understand contemporary environmental and global issues influencing the chemical and biological industries

(h) the broad education necessary to understand the impact of engineering solutions in a global, economic, environmental, and social context

The Department of Chemical and Paper Engineering measured metrics (performance indicators):

H1 Understand contemporary environmental and global issues influencing the chemical and biological industries

H2 Understand safety and regulatory issues related to the hazards associated with processes and the importance of ethical decisions related to these areas, real life examples

(j) a knowledge of contemporary issues

The Department of Chemical and Paper Engineering measured metrics (performance indicators):

J1 Conduct independent learning (outside classroom)

J2 Conduct international literature search using electronic library and utilize resources such as reference books, journal articles, and internet sources to analyze and design processes

By directly integrating performance indicators that assess global issues, the College of Engineering and Applied Sciences directly assesses the success of our globally-focused strategic vision and objectives by integrating this focus into ABET performance indicators. It is necessary and appropriate to build assessment of global engagement and context into ABET performance indicators, especially when student outcomes involve environmental, ethical, economic, safety, and contemporary knowledge, as the world is increasingly integrated with more and more issues being relevant across all countries. Furthermore, global awareness and education of our students will prepare them for this new integrated world and open doors for opportunity for collaboration and involvement.

As demonstrated, the College of Engineering and Applied Sciences has in place a well-established Knowledge, Skills, and Perspectives in the their Global Learning Outcomes as required for ABET accreditation and this MOA.

References

The Gold Standard 2010, Western Michigan University Strategic Plan, www.wmiche.edu

Students learning outcomes, abet.org

453451 Chemical Engineering Education for the Millennial Generation

Abstract:

The millennial generation are grown with the advance in technology and computation . They are the group of people who are faithful fans of Facebook, iPhone and iPad. They have also become the major talent group of chemical industry. Therefore the chemical engineering education for the millennial generation has become a very important topic in most of the countries. This presentation will provide you the challenge and opportunity of chemical engineering education for the millennial generation. Furthermore, the leadership of AIChE Young Professional Committee in this field will also be shared.

449538 More about the *Tendencies of Chemical Engineering in the Last 45 Years. What Is the Future?*

Abstract:

This contribution is a continuation of that of the AICHE 2015, and now the studied span of time was extended since 1970 to the date. The general statements of the 2015-material were that, according to the consulted sources, before the year 2000, chemical engineering received the beneficial of the advances in computation and internet, but in general terms, the traditional topics of chemical engineering were studied, such as: Chemical kinetics and reactors, catalysis and surface phenomena, unit operations, chemical thermodynamics, transport phenomena, simulation and optimization, chemical processes, process control, among others.

More or less after 2000, it is notorious that researchers started to have interest in new topics, where chemical engineering can have influence or can take advantage of the knowledge of other careers. Each time chemical engineers are eager and available to face and deal with topics from other careers and disciplines, and this fact makes chemical engineering stronger, due it extends its influence on other careers. Besides, other disciplines started to have interest in using the principles of chemical engineering and as a result, it is reinforcing and sharing its topics with other careers. In other words, the apparent walls and boundaries of chemical engineering with other careers started to fall or disappear.

The new lines, among others, studied in the last fifteen years or more have been: Environment, bioengineering, nuclear engineering, material science, thin films, renewable energies, food science, pharmacy, medicine, fuel cells, hydrogen power, nanotechnology, biotechnology, etc. Now, using the principles and theories of thermodynamics, transport phenomena, numerical methods, fluid dynamics, unit operations, chemical reactors, balances of mass, energy and entropy, etc, it is possible to have contribution in the purification of the environment, the production of alternative energies, food, artificial kidneys, etc.

In this 2016 - contribution, a deeper research was carried out, inasmuch as more information from internet was collected and analyzed, contents of more indexed reviews were searched, and this allowed distinguishing the tendency and the focus of each review.

The academic programs of chemical engineering of more universities in the five continents were consulted, and in this moment, although roughly, it is possible to confirm that from one geographic region to other, the focus of chemical engineering can change. The curriculum vitae of more researches were consulted; the research lines of more institutes were also investigated.

All of this allows saying that chemical Engineering has been changing and broadening its tendency and influencing in the last fifteen years in a significant form. Of course these new lines started before. Chemical Engineers, researchers, institutions, etc., are focusing their branches on the use of alternative feedstock such as biomass and waste to produce chemicals and energy. Their goal is to use green chemistry with less carbon in the industrial processes and sustainable processing.

The reasons of all these changes are because time is very dynamic, each time the social needs demand more energy, new food, different entertainment, new cloth, etc. The way to obtain and process the natural resources each time is different and requests more technologies and the interaction among the different disciplines must be enhanced because the industrial processes are each time more complex. Now the contamination must be avoided, there are stricter regulations about the environment. Search for new energies must be performed due hydrocarbons are going to be depleted. These external natural and social factors have been and will be the guide of the activities and research in chemical engineering for the next years.

476571 Pulse-Heating Method for Measuring the Critical Temperatures and Pressures of Thermally Unstable Compounds

Abstract:

The pulse-heating method for the measurement of the critical temperatures and pressures of thermally unstable compounds has been presented. The method is based on the measuring the temperature of the attainable superheat of a liquid. With increasing pressure, the temperature of the attainable superheat tends to the critical temperature. The temperature of the attainable superheat is measured with the help of a wire probe heated by electric current pulses, 0.01-1.0 ms in length. The short residence times provide little decomposition of compounds in the course of the experiments. The uncertainties of the critical properties measured are as follows: ,Â (is the absolute temperature) for â €œusualâ€ low-molar-mass compounds and ,Â for compounds with hydrogen bonding or with high acentric factor ().
The results of the measurements of the critical properties of about 200 compounds have been discussed.

466273 Current Status and Prospects of Industrial Xylitol Production in China

Abstract:

Xylitol is an alternative sweetener that has many beneficial health properties including regulating insulin-independent metabolism in humans, preventing dental caries and a number of pediatric diseases. Based on these promising advantages, xylitol has been commercially developed as a valuable ingredient for use in food and pharmaceutical industries. In China, xylitol is mainly produced industrially by the reduction of xylose derived from hemicellulose hydrolysates of biomass materials in the presence of Raney nickel catalysts. Currently, xylitol market lies between 20,000 and 40,000 tons per year with an economic value ranging between 90 and 340 million dollars and the xylitol market is expect to increase by 15% per year. The mainly Chinese companies including FUTASTE, LONGLIVE and HUAKANG PHARMA can produce about 40,000 tons per year. While chemical synthesis of xylitol has some disadvantages including a high energy requirement, extensive purification steps, and a high cost of product. Accordingly, the biological production of xylitol has recently attracted much attention. Many efforts have been made for the biological conversion of xylitol using microorganisms from hemicellulosic hydrolyzate by Chinese scientists. It is noteworthy that the xylitol productivity of recombinant *Escherichia coli* (1.84 $g\ l^{-1}\ h^{-1}$ and 143.8 $g\ l^{-1}$ yield) by our previous work and the xylitol productivity of recombinant *Candida tropicalis* (2.01 $g\ l^{-1}\ h^{-1}$ and 75.1 $g\ l^{-1}$yield) by Dr. Yuan's group are the highest xylitol productivity rates among recombinant bacteria and fungi from hemicellulosic hydrolyzate, respectively. Further, the industrial bio-production of xylitol and arabinose using *C.tropicalis* by Thomson Biotech from xylose mother liquor has been reported. All of above processes have been studied in an effort to provide one that is high yielding and competitive with chemical processes.

457047 Rational Substitution of Firefly Luciferase from *Photinus Pyralis* for Improvement of Thermostability

Abstract:

Firefly luciferase is widely used in many areas of biotechnology and molecular biology, but the application is restricted due to its poor thermal stability. Random mutagenesis and directed evolution paly an important role in the thermal stability of protein; nevertheless, large number of experimental attempts appear to be time- and labor-consuming. In this paper, to enhance the thermal stability of firefly luciferse, we got two mutants (E325K, E354K) by reversing the charges of individual residues at active sites with opposite interaction energies on the basis of molecular dynamic (MD) simulation. Root-mean-square-deviation(RMSD) and radius of gyration were used to evaluate the thermostability of the mutants. The residual activities of wide-type luciferase, E325K, E354K were 4.52%,7.22% and 30%, respectively, after being heated at 37 â„ƒ for 5 h, which showed the same trends with the computational predictions (E354K > E325K > WT). It was worth nothing that the T_{50} and T_m of E354K were 4.45 and 3.2 â„ƒ higher, respectively, than that of the native luciferase. Rational substitution of residue in regulation of thermostability using MD simulation would be a faster, less resource-intensive and more effective than random mutagenesis and directed evolution for instructional design to enhance thethermostability.

466981 Improving Cellular Robustness and Butanol Titers of *Clostridium Acetobutylicum* ATCC824 By Introducing the Heat Shock Proteins from Extremophilic Bacteria

Abstract:

In recent years, increasing concerns over environment, energy and climate have renewed interest in biotechnological production of butanol. However, growth inhibition by fermentation products and inhibitory components of raw biomass hindered the development of ABE fermentation. Improving cellular robustness of Clostridium acetobutylicum is of great importance for efï¬cient ABE production. In this study, we attempt to improve the robustness and butanol titers of C. acetobutylicum ATCC824 by overexpressing the groESL and dnaK from the extremely radioresistant bacteria Deinococcus wulumuqiensis R12 and C. acetobutylicum ATCC824. Three recombinants were obtained and designated as ATCC824(groESLR12), ATCC824(dnaK R12) and ATCC824(groESL824). These three recombinants were found to signiï¬cantly improve diverse types of stress-tolerances including furfural, butanol, oxidative, acid and osmotic-tolerance. Meanwhile, the butanol titers increased to 11.22g/L, 13.04g/L and 10.68g/L, which were 29.7%, 51% and 23.5% higher compared with the wild type, respectively. Especially for ATCC824(dnaK R12), the production of acetic and butyric acid decreased about 50% and 100% compared with the wild strain. These results indicated that overexpression of groESL and dnaK enhanced the metabolic rate of EMP pathway and improved acids re-uptake efficiency. In addition, ATCC824(groESLR12) was superior to ATCC824(groESL824) in diverse types of stress-tolerances and butanol titers because GroESL from an extreme bacterium could more efficiently perform their function in the heterologous host. Our study provided a new clue and a theoretical basis for improve the robustness and butanol titers of C. acetobutylicum ATCC824.

455883 Polymalic Acid Fermentation: A Platform for Production of Biopolymers and Chemicals

Abstract:

Polymalic acid (PMA) is a novel polyester polymer that has been broadly used in the medical and food industries. Its monomer, L-malic acid, is also a potential C4 platform chemical and extensively used in the food industry while it is currently supplied with the racemic mixture (DL-Malic acid) through chemical synthesis. Recently, great researches have been attempted to develop biotechnological process for L-malic acid production. In our study, a novel route for production of L-malic acid and its derivatives based on PMA platform was developed. Here, a specific fluorescence probe for detecting acidity strength was designed and used for high-throughput screening. Based on the genome-scale metabolic model *i*ZX637, Engineered the strain with overexpressed the gene of pyruvic carboxylase resulted in a PMA titer increase by 30% in shake flask. During the fermentation process, nitrogen limitation was found to be beneficial for PMA biosynthesis, which could upregulate the expression of key genes involved in the PMA biosynthesis pathway. Disruption or overexpression of the nitrogen transcriptional regulator Gat1 controlled by the TOR signaling pathway could obviously regulate PMA biosynthesis. Finally, the downstream process and progress in industry in China are presented.

476737 Engineering Yeast with Minicellulosome and Cellodextrin Pathway for Co-Utilization of Cellulose and Mixed Sugars

Abstract:

Consolidated bioprocessing (CBP), which integrates cellulase production, cellulose saccharification, and fermentation into one step, has been widely considered as the ultimate low-cost configuration for producing second-generation bioethanol from lignocellulose. However, the requirement of a microbial strain able to hydrolyze cellulosic biomass and convert the resulting sugars into high-titer ethanol limits CBP application. In this work, cellulolytic yeasts were developed by engineering ***Saccharomyces cerevisiae*** with a heterologous cellodextrin utilization pathway and bifunctional minicellulosomes. The cell-displayed minicellulosome was two-scaffoldin derived, and contained an endoglucanase and an exoglucanase, while the intracellular cellodextrin pathway consisted of a cellodextrin transporter and a β-glucosidase, which mimicked the unique cellulose-utilization system in ***Clostridium thermocellum*** and allowed ***S. cerevisiae*** to degrade and use cellulose without glucose inhibition/repression on cellulases and mixed-sugar uptake. Consequently, only a small inoculation of the non-induced yeast cells was required to efficiently co-convert both cellulose and galactose to ethanol in a single-step co-fermentation process, achieving a high specific productivity of ~62.61 mg cellulosic ethanol/g cell·h from carboxymethyl cellulose (CMC) and ~56.37 mg cellulosic ethanol/g cell·h from phosphoric acid-swollen cellulose (PASC). Our work provides a versatile engineering strategy for co-conversion of cellulose-mixed sugars to ethanol by ***S. cerevisiae***, and the achievements in this work may further promote the application of CBP in cellulosic biofuel production.

476734 Integrated Chemical and Bio-Catalysis forÂ Production ofÂ High-Value Liquid Transportation Fuels from Agricultural ResiduesÂ

Abstract:

Catalytic conversion of lignocellulosic biomass to high-valued transportation petrol, jet and diesel fuels is of great importance to develop versatile renewable energy and boost rural economy, thus receiving worldwide attention. However, the state-of-the-art processes still suffer from low efficiency in yield and productivity, which hinders the progress of its commercialization. Thus, it is essential to explore the consolidated catalytic reaction and refinery process. Here we report an advanced approach/process that could continuously produce long-chain (C5-C15) ketones from lignocellulosic biomass in a cascade mode by integrating several bio- and chemical catalysis reactions, which exhibits high efficiency and stable conversion rate (>70 %) in alkylation reaction. This technology consolidated the acetone-butanol-ethanol (ABE) fermentation, palladium catalyzed alkylation and in situproduct recovery, allowing the catalytic reaction to proceed consistently and smoothly. The remarkable conversion property is mainly attributed to excellent compatible linkage of chemical catalysis with biosynthesis and process design to stabilize the activity of Pd catalyst. This validated strategy for rational process design and integration of chemical- and bio-catalysis offers great demonstration in producing high-valued transportation biofuels derived from agricultural residues.

457314 Cholesteryl Hyaluronic Acid-Coated, Reduced Graphene Oxidenanosheets for Anti-Cancer Drug Delivery

Abstract:

Here, we report hyaluronyl reduced graphene oxide (rGO) nanosheets as a tumor-targeting delivery system for anticancer agents. Hyaluronyl-modified rGO nanosheets were prepared by synthesizing cholesteryl hyaluronic acid (CHA) and using it to coat rGO nanosheets, yielding CHA-rGO. Compared with rGO, CHA-rGO nanosheets showed increased colloidal stability under physiological conditions and improved in vivo safety, with a survival rate of 100% after intravenous administration of 40 mg/kg in mice. The doxorubicin (Dox) loading capacity of CHA-rGO was 4-fold greater than that of rGO. Uptake of Dox by CD44-overexpressing KB cells was higher for CHA-rGO than for rGO, and was decreased in the presence of hyaluronic acid through competition for CD44 receptor binding. After intravenous administration in tumor-bearing mice, CHA-rGO/Dox showed higher tumor accumulation than rGO/Dox. The in vivo antitumor efficacy of Dox delivered by CHA-rGO was significantly increased compared with free Dox or rGO/Dox. In CHA-rGO/Dox-treated mice, tumor weights were reduced to 14.1% 0.1% of those in untreated mice. Our findings indicate that CHA-rGO nanosheets possess greater stability, safety, drug loading capacity, and CD44-mediated delivery of Dox than rGO nanosheets. These beneficial properties of CHA-rGO improved the distribution of Dox to tumors and facilitated the cellular uptake of Dox by CD44-overexpressing tumor cells, resulting in enhanced anticancer effects.

AUTHOR INDEX

AUTHOR INDEX

AIChE
120 Wall Street, FL 23
New York, NY 10005-4020

ISBN 978-1-5108-3460-6

CURRAN ASSOCIATES INC.
proceedings.com